Carbon Coalitions

Carbon Coalitions

Business, Climate Politics, and the Rise of Emissions Trading

Jonas Meckling

The MIT Press
Cambridge, Massachusetts
London, England

For information about special quantity discounts, please e-mail special_sales@mitpress.mit.edu.

This book was set in Sabon by Graphic Composition, Inc., Bogart, Georgia. Printed and bound in the United States of America.

Library of Congress Cataloging-in-Publication Data

Meckling, Jonas.
Carbon coalitions : business, climate politics, and the rise of emissions trading / Jonas Meckling.
p. cm.
Includes bibliographical references and index.
ISBN 978-0-262-01632-2 (hardcover : alk. paper) — ISBN 978-0-262-51633-4 (pbk. : alk. paper)
1. Emissions trading. 2. Climatic changes—Government policy. 3. Emissions trading—United States. 4. Emissions trading—European Union countries. 5. Emissions trading—Great Britian. 6. Climatic changes—Government policy—United States. 7. Climatic changes—Government policy—European Union countries. 8. Climatic changes—Government policy—Great Britain. I. Title.
HC79.P55M43 2011
363.738'746—dc22

2011003547

10 9 8 7 6 5 4 3 2 1

For the fantastic four:
Doris, Hans, Karen, and Anna

Contents

Acknowledgments

A book is like an iceberg: only the smaller part of the whole is visible. This book rests on the contributions of a great many people to whom I owe thanks for their support, encouragement, and constructive criticism. Robert Falkner, my doctoral adviser at the London School of Economics and Political Science, merits special appreciation, as he was an excellent guide throughout my doctorate. As much as the topic of this book spans the Atlantic, so did the actual research process. I wrote the larger part of the book during my fellowships at the Belfer Center for Science and International Affairs, and the Mossavar-Rahmani Center for Business and Government at Harvard University. Kelly Sims Gallagher, John Holdren, and Henry Lee offered a most inspiring intellectual home in the Energy Technology Innovation Policy group as well as many valuable insights into international and U.S. climate politics. Robert Stavins and Robert Stowe of the Harvard Project on Climate Agreements gave me the wonderful opportunity to revise the thesis for publication as a book. I am grateful for their hospitality and generous support.

Over the years, the work resulting in this book was made possible through funding from the German National Academic Foundation, the Consortium on Energy Policy Research at Harvard, the Environment and Natural Resources Program at the Harvard Kennedy School, the Ev. Studienwerk Villigst e.V., and the Fritz Thyssen Foundation. Funding through the McCloy Fellowship in Environmental Affairs of the American Council on Germany allowed for extensive fieldwork in the United States.

This book would not have been possible without those policymakers, activists, and lobbyists in Europe and the United States who were willing to take time out of their busy schedules for interviews. I was often struck by the commitment they brought to the interviews as well. They have been an essential source for original insights into the globalization of carbon trading. A great number of other people helped along the way—only some of which I can name here. The effectiveness of the interview phase in Washington, DC, was tremendously enhanced by Mark Starik's kind offer of an office at the Department of Strategic Management and Public Policy at the George Washington University Business School. Special thanks also go to Ans Kolk and David Levy, who gave me access to their data collection on the role of oil majors in climate politics in the 1990s, thereby providing an additional critical source of information. I am indebted to Stacy VanDeveer, too, for guiding me through the various steps of publishing a book. I am grateful for the extensive and valuable comments on the manuscript from Frank Biermann, Stephen Woolcock, and three anonymous reviewers. Finally, I owe thanks to Clay Morgan and his colleagues from the MIT Press who carefully shepherded me through the process of preparing the manuscript for publication.

This book would never have been written had it not been for friends and family keeping me grounded, lending moral support, and sharing laughter and good times. In all my years of academic soul-searching, my parents, Doris and Hans Meckling, have been extraordinarily supportive, instilling in me the belief that it is a worthwhile journey to pursue my interests and passions. To them and my sisters, Karen and Anna, I dedicate this work.

1 Introduction

On May 19, 1997, John Browne, then CEO of British Petroleum (BP), entered the stage at Stanford Business School to deliver what would become a landmark speech. Breaking ranks with his industry peers who opposed emissions controls, Browne said that the time for action on global climate change had come—and that BP was working with the U.S.-based group Environmental Defense to establish an internal emissions trading system to meet its own greenhouse gas (GHG) emissions reduction target.[1] BP launched its in-house pilot trading scheme one year later. And by 2000, that pilot grew to encompass the entire organization, effectively establishing the first international GHG emissions trading scheme.

A decade into the new century, BP's internal experiment has become the global experiment of climate politics: by generating and trading the right to emit GHG emissions, firms, governments, and individuals aim to slow, stop, and eventually reverse the worldwide growth in emissions. Originally developed in the United States, carbon trading has emerged as the industrialized world's primary response to global climate change. A number of so-called son-of-Kyoto bills were passed over the past decade or are under consideration in developed countries, including the European Union (EU) and United States. In addition, parts of the developing world—in particular China, India, and Brazil—are involved in the global trade of carbon dioxide (CO_2) emissions permits through the generation of project-based credits. The regulatory approach, in short, has gained significant political momentum, leading to the emergence of new commodity markets: the carbon markets. In 2009, the carbon markets were worth US$144 billion, and they are expected to grow exponentially over the coming decade (Kossoy and Ambrosi 2010).

The creation of new markets to solve an environmental problem is a new and striking phenomenon in global environmental politics. It represents the most large-scale manifestation of a broader trend toward market-based global environmental governance. Since the early 1990s, the regulatory style of environmental policy has shifted markedly from command-and-control regulation to market-based forms of governance: "The Invisible Green Hand" (2002) has come to rule global environmental affairs. This trend follows the idea that the market is "the source of innovation, efficiency, and incentives necessary to combat environmental degradation without compromising economic growth" (Newell 2008a, 79). Market-based environmental governance mainly comes in two ways: the growth of private governance mechanisms, and the creation of property rights in regard to environmental goods (see Falkner 2003; Newell 2008a; Pattberg and Stripple 2008). While private governance includes voluntary agreements, ecolabels, and public-private partnerships, assigning property rights refers to ecotaxes and, most notably, emissions-trading schemes. When it comes to the scale and geographic scope of market-based environmental governance, carbon trading is the most significant case. The policy instrument has emerged as the central pillar of climate policies. In this process it outcompeted carbon taxes, regulatory standards, and voluntary climate policy. While these still play a role in climate policy mixes, they are not the major regulatory instruments.

The worldwide spread of emissions trading presents a puzzle. The approach has been a highly controversial policy instrument. On the one hand, supporters of market-based climate governance stressed its efficiency and argued that it unleashes the potential of the private sector in solving the climate change problem (Holliday, Schmidheiny, and Watts 2002). On the other hand, critics claimed that emissions trading grants firms a *license to pollute*, leading to a weakening of environmental regulation (Corporate Europe Observatory 2000; Bachram 2004; Lohmann 2006). Governments and green groups in the EU were initially opposed to the use of market mechanisms in climate politics. Yet the EU came to pioneer GHG emissions trading by developing the first cross-border trading scheme, the EU Emission Trading Scheme (EU ETS).

Interestingly, the rise of carbon trading in global climate politics coincided with a shift in the political strategy of a number of influential firms from opposing climate regulation to advocating for a specific policy solution. There is considerable evidence that the outcome of international climate politics in general and the rise of carbon markets in particular have been strongly affected by the political behavior of firms. Corporate activists such as BP and DuPont started to advocate emissions trading as a policy response to climate change in the mid-1990s, assuming a leading role in popularizing the policy instrument across the Organization for Economic Cooperation and Development world.

This book sets out to explore the strategies, level of influence, and sources of influence of business with regard to market-based climate policy. It asks two questions: What role did business play in the global rise of emissions trading? And why did firms succeed in promoting emissions trading as the main policy solution to global climate change? In the following section, I lay out the key argument on the role and influence of business in the rise of emissions trading.

Business Coalitions in Global Climate Politics

The research questions speak to the debate on the power of business in global environmental politics. How powerful are firms in shaping the agenda and outcome of global environmental politics? What are their sources of influence? Over the last three decades, business has emerged as a critical actor in global environmental politics in several ways. Business actors lobby parties to international negotiations (Coen 2005; Fuchs 2005a), shape public discourses (Sell and Prakash 2004), and provide technological solutions to environmental problems (Falkner 2005); they are involved in rule setting as well (Falkner 2003; Pattberg 2007). While the economic preponderance of firms bestows them with a competitive advantage vis-à-vis other interest groups, corporate power also faces a number of constraints (Falkner 2008). First, countervailing forces, such as states and environmental groups, limit the power of business. Their relative political weight and the level of contestation on a policy issue affect the influence of business. In addition, divisions in the business community cause conflict between and within business sectors, which constrains the overall political clout of business as such (Cox 1996b; Skidmore-Hess 1996). The distributional effects of environmental policy, and regulatory policy in general, lead sectors and firms to different policy preferences, which can result in inter- or intraindustry conflict. After all, "political competition follows in the wake of economic competition" (Epstein 1969, 142).

Given these limitations to their influence, firms, I contend, face the challenge of organizing collective action to achieve political clout. As Philip Cerny (2003, 156) says, political outcomes "are determined not by simple coercion and/or structural power but, even more significantly, by how coalitions and networks are built in real-time conditions among a plurality of actors" (see also Mattli and Woods 2009). Collective action, in other words, becomes a source of "power through organization" (Offe and Wiesenthal 1980, 72), as it allows firms to pool political resources and pursue effective collective strategies. These arguments reflect a pluralist/neopluralist understanding of world politics (Mattli and Woods 2009; Avant, Finnemore, and Sell 2010; Cerny 2010), in which this book is firmly embedded. The central assumption is that individual and collective actors "engage in processes of conflict, competition, and coalition building in order to pursue those interests" (Cerny 2010, 4). With regard to the role of business in global environmental politics, Robert Falkner (2008) has developed a neopluralist approach, stressing the possibility of business conflict. This book builds on the neopluralist literature by advancing our understanding of transnational business coalitions as a source of corporate power. The role of firms in global environmental politics needs to be understood in terms of both business conflict and cooperation.

Through this lens, international climate policy appears—to a large extent—as the product of shifting and competing business coalitions along with their respective influence. Hence the title of this book: *Carbon Coalitions*. I argue that business could not prevent mandatory emissions controls but instead succeeded in influencing the regulatory approach in favor of market-based climate policy. In the early phase of international climate politics, business stood largely united in opposition to caps on GHG emissions. Dominated by U.S. fossil fuel interests, the antiregulatory coalition successfully vetoed international and domestic emissions controls (Levy and Egan 1998; Goel 2004). In particular, it fended off proposals for carbon/energy taxes in both the EU and the United States, effectively demonstrating de facto veto power with regard to environmental taxes.

This dynamic started to shift in the run-up to the Kyoto conference in the mid-1990s, as a split in the business community emerged. A number of firms based in the United Kingdom and the United States began to promote emissions trading in tandem with business-oriented nongovernmental organizations (NGOs) and a few public actors (Yandle 1998; Yandle and Buck 2002; Matthews and Paterson 2005). This effort led to the emergence of a protrading NGO-business coalition. At its core stood the political goal of promoting emissions trading as the policy response of choice to climate change. The coalition advocated carbon trading, shared information, and generated market infrastructure. While transnational in scope, the coalition crystallized in more local advocacy coalitions at regional, national, and subnational levels.

By joining forces behind a market-based climate policy, the protrading lobbies among business and NGOs sidelined more radical business and environmental coalitions that favored no emissions controls, or carbon taxes and command-and-control policies, respectively. The protrading coalition came to define a climate compromise. Business supporters of emissions trading were initially driven by an "antitaxation" agenda that aimed to prevent the implementation of carbon taxes, which were perceived as a more costly form of regulation compared to carbon trading. Big emitters from the oil industry and especially the power sector were among the first movers. By showing policy leadership on emissions trading and legitimizing mandatory yet market-based climate policy, business created an opportunity for governments to push for binding emissions reduction targets and market mechanisms. While business did not determine political outcomes, it has proven to have considerable influence on regulatory style, effectively defining the range of available policy instruments. Without the emergence of transnational business support for the approach, carbon trading would have been unlikely to globalize within a decade.

Why did the protrading coalition succeed in grafting carbon trading on to the agenda of international climate politics? I argue that its success is a function of political opportunity structures, the political resources that the coalition could leverage, and the strategies it pursued. By considering these three elements—political environment, power resources, and strategy—this approach advances existing studies of the role of business in global environmental politics (Levy and Newell 2005; Falkner 2008; Clapp and Fuchs 2009a). Adopting a power-oriented framework, these studies predominantly focus on different forms of corporate power, such as instrumental, discursive, and structural power, thereby establishing a multidimensional understanding of corporate power. This book goes one step further by acknowledging that—next to power resources—the political environment (Cerny 2010) and political strategy (Levy and Scully 2007) matter significantly in the equation of business influence. It thus adopts an influence-oriented perspective as opposed to a power-oriented one (see Betsill and Corell 2008).

In terms of political opportunities, policy crises and international and/or domestic norms are key. The protrading coalition could capitalize on the political stalemate between environmental interests and economics interests, and between the EU and the United States, by suggesting a third policy solution: a mandatory yet market-based climate regime. In addition, carbon trading demonstrated a good fit with the liberal norms embedded in the system of global environmental governance. While political opportunities were critical, the coalition's resources and strategies allowed the protrading advocates to exploit such circumstances. Coalitions allow firms to pool material and immaterial resources—notably money and legitimacy. Effective coalitions also are able to pursue strategies that individual firms could hardly achieve.

Two strategies proved to be particularly crucial: mobilizing state allies, and playing multilevel games. The success of the protrading coalition relied on powerful state allies such as the U.S. administration and the European Commission. Given the key role played by state allies, the diffusion of emissions trading essentially represents a case of multileadership of factions of firms and market-oriented government actors. A second important strategy of the transnational coalition was that it targeted multiple political levels. It influenced policy processes at the subnational, national, regional, and international levels, which led to a broad shift in the global climate agenda. A single company could most likely not have achieved this. The confluence of these factors allowed the protrading coalition to outcompete other advocacy groups in influencing the regulatory style of climate policy. The transnational coalition became a vehicle for the spread of a U.S. regulatory approach. While I put forward a coalition-based, business-centered reading of the rise of carbon trading, the approach faces competition from another business-centered approach as well as more general international relations theories.

Alternative Explanations: Capital, States, and Ideas

The global rise of carbon trading can be viewed through a number of alternative theoretical lenses—notably neo-Gramscianism, neoliberal institutionalism, and constructivism. While neo-Gramscians offer an alternative business-centered reading with a focus on the structural power of global capital, neoliberal institutionalism and constructivism represent standard models of international environmental cooperation. Each of the three theoretical perspectives helps to explain different aspects of the spread of emissions trading. In what follows, I will discuss the contributions and limitations of these alternative theoretical views. While this is not the place to rehearse the entire literature on the different theoretical traditions, I will look at a few key aspects that are relevant to the empirical puzzle. It should be noted that the exploration here does not exhaust the spectrum of potential theoretical perspectives on the globalization of carbon trading. Neorealism and hegemonic stability theory would be, for instance, alternative lenses. Yet these approaches have existed only at the margins of the academic field of global environmental politics, as they have contributed relatively little to explaining international environmental cooperation. I therefore will not include them in the discussion below.

Capital-Centered Explanations

Neo-Gramscian thought in international political economy (Gill and Law 1989; Cox and Sinclair 1996) has inspired a number of scholars working on the power of firms in global environmental politics (Newell and Paterson 1998; Levy and Newell 2002; Levy and Egan 2003; Levy and Scully 2007). Their efforts have led to a well-developed body of literature. Given its analytic focus on corporate power, the neo-Gramscian approach is the main reference point for an examination of the neopluralist perspective. Both represent business-centered approaches.

The conceptual cornerstone of the neo-Gramscian approach is global hegemony, or a historical bloc, which is to be understood as resting "on a specific configuration of societal groups, economic structures, and concomitant ideological superstructures. A historical bloc exercises hegemony through the coercive and bureaucratic authority of the state, dominance in the economic realm, and the consensual legitimacy of civil society" (Levy and Egan 2003, 806). Hence, hegemony is an alignment of political, economic, and ideological forces that coordinates the behavior of major social groups. Neo-Gramscian notions of a transnational historical bloc depart from classical Marxism insofar as social forces in a "war of position," in which the structural power of business has to be continuously reinstated, are thought to contest the dominant state-business-civil society alliance (ibid.). These power struggles often result in the reconfiguration of the historical bloc because capital has to accommodate social pressure and redefine its general interest. In short, the neo-Gramscian approach puts forward the notion of contested business hegemony.

The notion of hegemony builds on a concept of the structural power of capital in the global economy. This power is derived from the tax revenue and employment provided by the critical economic sectors on which the state depends (Gill and Law 1989; Gill 1993). While the neo-Gramscian approach considers different forms of corporate power, it implicitly assumes a hierarchy among them. Structural power is the power of first order, from which all other forms of power are derived. Though this is rarely made explicit, the assumption that political processes gravitate toward the creation of historical blocs, which reflect the capital's interest, suggests that structural power maintains the status of primus inter pares among the different forms of power. The concept of hegemony, however, does not imply that capital, or business, is a homogeneous group with a single interest. Neo-Gramscians are well aware that different factions of business can have diverging interests because sectors and firms are differently affected by regulation (Cox 1996b; Skidmore-Hess 1996). Rather, the assertion is that core sectors in the economy represent "capital-in-general" (Newell and Paterson 1998). These are the strategic sectors that the entire capitalist system builds on, such as energy industries or finance—the so-called commanding heights. Accordingly, the historical bloc reflects the interest of those sectors representing capital-in-general. The role of the state in capital accumulation is to identify and advance the general interest of capital. In this process, the state itself internationalizes by creating multilateral and supranational institutions to promote as well as reproduce the global economy (Jessop 1990; Cox and Sinclair 1996). The neo-Gramscian approach and my focus on the role of state allies as a source of corporate power are in this respect similar.

With regard to climate politics, neo-Gramscians suggest a reconfiguration of the transnational historical bloc (Levy and Egan 2003). In the pre-Kyoto period, the fossil fuel and energy-intensive manufacturing industries were at the core of a historical bloc. They represented global capital, some scholars contend, due to the historically close link between energy production and economic growth (Newell and Paterson 1998; Levy and Egan 2003). The lobbies of these sectors, foremost the Global Climate Coalition, successfully prevented mandatory GHG controls for a long time (Levy and Egan 1998). In short, the fossil fuel industries and their state allies maintained an antiregulatory historical bloc. This changed when a rift in the business community appeared prior to the Kyoto conference in 1997. David Levy and Daniel Egan (2003) assume a reconfiguration of the transnational historical bloc. In response to miscalculations in political strategy and early regulation in Europe, which created prospects for markets for low-carbon technologies and made European firms move ahead, U.S. companies began to change strategy. "The climate regime associated with this bloc provides very limited targets for emissions reductions, market-based implementation mechanisms, and minimal regulatory intrusion upon corporate autonomy" (ibid., 818).

The neo-Gramscian concept of contested hegemony and reconfiguration appears to offer a powerful explanation for strategic change in the business community, as it encompasses political, economic, and ideological phenomena. As far as the description of corporate political activity is concerned, neo-Gramscian thought has significantly advanced the study of the firm's political economy. This is especially true for the ideational dimensions of business activity, as neo-Gramscians have brought the discursive dimension of business power into the analysis (Levy and Egan 1998; Levy and Newell 2002; Fuchs 2005b). Moreover, the idea of the contestedness of business reflects on the role of conflict in the relations of firms with other actors, such as governments and civil society groups. This goes beyond notions of the embeddedness of firms, which stress that firms simply respond to institutional pressure (Granovetter 1985; Sally 1994).

Yet when it comes to an explanation, neo-Gramscian concepts of "contested hegemony" raise questions as to their internal coherence. While considering political struggles as the norm in capitalist states, neo-Gramscians also assume the existence of a historical bloc. The historical bloc is hegemonic, in short, but it emerges in a nonhegemonic way. This remains the central puzzle of the neo-Gramscian approach: Why should a bloc emerge if social contests dominate? Or why should pluralist struggles matter at all if the structural power of capital prevails? To be clear: neo-Gramscians pay close attention to the formation of historical blocs and alliance building (Levy and Newell 2002; Matthews and Paterson 2005; Newell 2008b). Hence, in terms of an empirical analysis, the neo-Gramscian and coalition-centered approaches advanced here are similar, as they both look at coalition building. But coming from a pluralist angle, I do not assume that the structural power of capital will ultimately lead to a dominant position of business. Neo-Gramscians instead assume that any war of position will gravitate toward a dominant, albeit reconfigured, historical bloc. The agency-structure debate thus is implicitly resolved in favor of the structural argument.

While both the neo-Gramscian approach and coalition-centered perspective advanced in this book stress the role of state allies as sources of business power, neo-Gramscians run the risk of overstating the alignment of state and business strategies. If the state was structurally dependent on business to the extent that neo-Gramscian approaches suggest, one could claim that there would be no need for lobbying (Falkner 2000). Yet firms invest significantly in lobbying activities to influence policy, as state allies cannot be taken for granted. Given this, neo-Gramscian interpretations tend to overemphasize the stability of the political order at the expense of acknowledging the contingencies of the actual political process. To use Alexander Wendt's words (1987, 362): "*Structural* analysis explains the possible; *historical* analysis explains the actual."

The neo-Gramscian approach, to wrap up, represents an appealing theoretical lens to study corporate power in global environmental politics due to its encompassing narrative of capitalist hegemony. It has critically advanced the study of business in global environmental politics. Much of the alliance-focused analysis of this book fits in well with a neo-Gramscian interpretation. Nevertheless, the key difference is that the analysis offered here does not have an analytic bias in favor of the structural power of business. It instead explores the actual influence of business in contingent political processes. In the conclusion, I will discuss the neo-Gramscian interpretation of the rise of market-based climate policy in light of the empirical findings.

State-Centered Explanations

Neoliberal institutionalism has emerged as the mainstream explanation for international environmental cooperation (Young 1989; Haas, Keohane, and Levy 1993). The assumption is that states cooperate to maximize absolute gains in areas where mutual benefits exist. In these cases, states create institutional arrangements, or regimes. Two variants of neoliberal institutionalism have played an important role in the study of global environmental politics: the contractarian strand, and the constitutive one (Young 1997). Contractarians assume that state identities and interests are exogenously given. Based on set interests, states negotiate cooperative agreements in situations where a collective strategy maximizes the joint gains. In international environmental politics, the transboundary nature of global environmental problems creates situations where rational state actors can mutually benefit through cooperation. The constitutive strand instead argues that international institutions, such as international agencies, as well as treaties and informal practices are not only the outcome of interstate negotiations but also facilitate intergovernmental cooperation. These institutions in fact shape state identities and interests.

Regarding international climate politics, neoliberal institutionalists suggest that the Kyoto Protocol serves the interests of its parties (Rowlands 2001). Emissions trading was included not because of the overriding power of the United States but because the United States and the EU struck a deal: while the EU accepted market mechanisms, the United States agreed to a binding target. From the perspective of the constitutive strand of neoliberal institutionalism, the UN Framework Convention on Climate Change (UNFCCC) facilitated the protocol's negotiation. As a preexisting international institution, this convention facilitated the bargaining process. The spread of emissions trading as a policy instrument could be interpreted as the implementation of an international regime at the national level. It would thus be a case of state compliance. An interest-based, statecentric perspective on international environmental cooperation takes into account that much of international environmental negotiation is essentially interest-based bargaining (see Barrett 2005). This approach also stresses that states are the primary actors in international climate cooperation, including the case of carbon trading. As I hope to demonstrate, business requires strong state allies to advance its agenda.

While neoliberal institutionalism offers complementary insights into the rise of emissions trading, it also shows blind spots regarding state interest formation and the content of international cooperation. First, in the classic variant of the contractarian perspective, state interests are exogenous, leaving their formation unexplained. Robert Keohane (1993, 285) eloquently makes this point: "In the absence of a specification of interests . . . institutionalist predictions about cooperation are indeterminate. That is, institutional theory takes states' conceptions of their interests as exogenous: unexplained within the terms of the theory." This explanatory void is due partly to the fact that nonstate actors are not an essential part of the neoliberal equation. While neoliberal institutionalists acknowledge the existence of nonstate actors (Keohane and Nye 1972), their analytic focus remains essentially statecentric. Accordingly, classic neoliberal institutionalism cannot explain why the United States pushed for the inclusion of market mechanisms in the Kyoto Protocol or why the EU shifted its position from strong opposition to strong support for emissions trading.

A number of institutionalist scholars have taken up this criticism, though, addressing the question of state interest formation in the context of neoliberal institutionalism (Milner 1997; Simmons 1997). They have done so by bringing domestic politics back into the equation, particularly by looking at the interaction of domestic actors and institutions. This strand of research converges with the neopluralist argument advanced here to the extent that it considers the role of interest groups in state interest formation. Neopluralism can essentially be understood as a theory of state interest formation with a primary focus on the role of business in shaping foreign policy. Yet the analysis in this book goes beyond the role of domestic firms in foreign policymaking by also considering the transnational organization of business and the engagement of business with international politics.

Second, neoliberal institutionalism does not make explicit assumptions about the content of international agreements. It is assumed that parties agreed on the optimal policy solution. Hence, it remains puzzling from an institutionalist perspective why states agreed on market mechanisms as opposed to competing policy instruments. Building on the rationalist assumptions of the approach, it could be asserted by theoretical extrapolation that emissions trading became the dominant international response to global climate change because it represents the most efficient solution for all actors involved. While many political actors have often portrayed international emissions trading as the most cost-efficient policy solution, a number of economists actually question the efficiency of emissions trading. Neoliberal institutionalism leaves aside the notion that the choice of policy instruments is essentially a political battle over competing norms, as opposed to a search for the optimal global policy solution. The narrative presented here very much reflects that emissions trading did not simply present itself as the most efficient solution. Rather, it took a transnational campaign to persuade key actors to accept carbon trading.

In sum, contractarian neoliberal institutionalism retains explanatory power as regards the question of why agreements are reached and the premise that states are the final decision makers in policy choices. These insights will be reflected in the business-centered approach advanced here by considering state allies as a key source of business influence. Classic neoliberal institutionalism remains problematic as an explanation of the rise of emissions trading, as it does not open the black box of state interests. Still, more recent developments of neoliberal institutionalism accounting for the role of domestic politics in the formation of state preferences align with much of the neopluralist argument. I will return to the discussion of neoliberal institutionalism in the book's concluding chapter against the backdrop of the empirical analysis.

Ideational Explanations

Constructivism offers a third perspective on the rise of carbon trading: "Constructivists focus on the role of ideas, norms, knowledge, culture, and argument in politics, stressing in particular the role of collectively held or *intersubjective* ideas and understandings on social life" (Finnemore and Sikkink 2001, 392). As such, a central tenet of the approach is the idea that political action is driven primarily by ideational instead of material factors (Ruggie 1998; Wendt 1999). Ideas and norms are the structures that shape interests and identities. Constructivism emerged in the context of critiques of statecentric approaches that concentrate on material power (Finnemore and Sikkink 1998). It draws on older theoretical traditions such as the English school and the sociological strand of institutionalism (Finnemore 1996; Checkel 1998).

The role of ideas and norms in facilitating international environmental cooperation has been considered in a number of ways (Haas 1992a Litfin 1995; Keck and Sikkink 1998; Bernstein 2001). Peter Haas (1992b) for instance, introduced the concept of epistemic communities, which are expert communities that build consensual knowledge about causal relationships. He argues that scientific knowledge about environmental problems is crucial to bringing about interstate cooperation. The Montreal Protocol on Substances that Deplete the Ozone Layer is a case in point, Haas says. While Haas focuses on causal beliefs, Margaret Keck and Kathryn Sikkink (1998) stress the role of principled, or normative, beliefs for cooperation to occur. They suggest that transnational environmental advocacy networks are vehicles for such norms in particular in issue areas such as the environment or human rights. Much of the constructivist notion of ideas in global politics is reflected by recent research on policy diffusion. This relatively new body of literature considers diffusion as a distinct social mechanism that builds on uncoordinated but interdependent decision making (Simmons and Elkins 2004; Busch and Jörgens 2005; Levi-Faur 2005). Such mechanism-oriented diffusion studies converge with constructivist notions of ideas and norms as ideational structures.

With regard to the marketization and liberalization of global environmental politics, Steven Bernstein (2001) has offered a powerful idea-based account in *The Compromise of Liberal Environmentalism*. His argument is that liberal—that is, market-based—environmental policies are the best fit with the underlying normative structure of a liberal international system. Bernstein sees global environmental politics as deeply embedded in liberal norms since its inception. The emergence of private governance and environmental markets are the visible reflections of the dominant idea of liberal environmentalism. Bernstein thus provides a powerful narrative for the marketization of environmental politics in general and the rise of carbon trading in particular. Concerning the latter, Bernstein (ibid., 118) writes: "Perhaps no better example of the effects of liberal environmentalism exists than the signing of the 1997 Kyoto Protocol. . . . [T]he Kyoto Protocol is the most ambitious attempt to date to implement market and other economic mechanisms at the global level that I have identified as a key component of liberal environmentalism." At first sight, the constructivist approach to international environmental cooperation appears to offer a powerful explanation. Both causal and normative beliefs play an important role in the process. Because the problems to be solved relate to the natural world, there is a high demand for scientific knowledge. The political goal of environmental protection also is highly contested and competes with other norms, such as economic growth and development. Constructivism has a strong appeal because it can explain the content of international cooperation.

The problem with structural constructivist explanations such as Bernstein's is that they fall short of explaining norm emergence and why certain ideas prevail over others. A significant, more agency-oriented strand in constructivism has tried to rectify this flaw by considering specific actors as vehicles of ideas. Epistemic communities (Haas 1992b) and transnational advocacy networks (Keck and Sikkink 1998; Risse, Ropp, and Sikkink 1999) are norm entrepreneurs that play an instrumental role in the emergence along with diffusion of ideas and norms.

This book builds on this agency-oriented strand of constructivist research, considering the role of business coalitions in struggles over ideas. The approach presented here goes one step further, however, by explicitly studying the capabilities that enable nonstate actors to spread ideas, thereby dealing with questions of power and influence. The existing agency-oriented perspectives of constructivism do not encompass whether ideas are driven by transnational nonstate actors, or whether transnational actors merely act as carriers of ideas in a quasiautomatic process of diffusion. As Jeffrey Checkel (1998) argues, even agency-oriented variants of constructivism lack an explicit theory of agency. My book clearly takes an agency-oriented standpoint, suggesting that ideas come to play a part in world politics only through the active promotion of agents. I view ideas as intervening variables between agents and political outcomes, not as causal variables. The spread of an idea is not quasiautomatic but rather depends critically on actors employing their power resources and skills in the process of "strategic social construction" (Finnemore and Sikkink 1998, 909).

This brings us to the question of the sources of power. Given its focus on ideas, constructivist analysis leans heavily toward ideational power structures at the cost of neglecting material forms of power. It can be argued, though, that interests and material forms of power often stand behind particular ideas (see Hall 1989). Material power, such as funding, is necessary to effectively advocate and disseminate an idea in a campaign. Yet the material foundation of the successful diffusion of ideas escapes the analytic lens of constructivists, because ideas are understood to be causal in themselves. In sum, constructivism offers highly valuable insights into the role of ideas in global environmental politics. Especially the agency-oriented strand in constructivism supplies helpful vantage points for developing an understanding of the role of business in political struggles over meaning. While making clear links to the constructivist research program, this study also departs from it insofar as it considers ideas as intervening variables driven by actors possessing both material and ideational power resources, as will be spelled out in chapter 2.

To conclude, this book positions itself in the debate on the power of business in global environmental politics. The arguments made here are embedded in a neopluralist understanding of business power, advancing our understanding of the organizational dimension of corporate power. This book thus complements the business conflict lens of existing neopluralist approaches to the role of business in global environmental politics with a coalition-centered analytic lens. Hence, the conceptual work on transnational business coalitions in this book is seen as a contribution to neopluralist thinking on business in global environmental politics in particular (Falkner 2008) and neopluralist approaches to world politics in general (Avant, Finnemore, and Sell 2010; Cerny 2010). The main advantage of a neopluralist perspective on the globalization of carbon trading lies in its high degree of analytic sensitivity toward nonstate actors as well as actual historical patterns of conflict and cooperation across state and nonstate actors. Neo-Gramscian scholars offer an alternative reading of the role of business in global climate politics. Beyond business-centered perspectives, neoliberal institutionalism and constructivism represent competing explanatory approaches.

All three contending approaches grasp important aspects of the diffusion of emissions trading. For instance, neoliberal institutionalism points to the crucial role of nation-states in setting the rules of carbon trading. Neo-Gramscian and constructivist analyses direct attention to the discursive processes around carbon trading. Moreover, the neopluralist reading of this book overlaps partly with certain aspects of the three alternative explanatory approaches. The narrative laid out here resonates with the interest of neo-Gramscians in alliance formation, the domestic politics approach in neoliberal institutionalism, and the concept of transnational advocacy networks in constructivism. For theoretical reasons relating to interest formation, assumptions about agency and structure, and the role of material power, however, this book advances a coalition-centered, pluralist perspective. It is argued that such a perspective sheds the most light on the actual historical role of business in the globalization of carbon trading. In the concluding chapter, I will return to this contention in relation to the empirical findings.

Methodological Issues

I now turn to this study's research design, which establishes the formal link between the theoretical argument and the empirical research. First, I operationalize the concept of influence and explain the methods used to infer causality. Next I discuss the case study design and selection of the cases. In a final step, I explain the primary and secondary sources of data along with the process of empirical research.

Methods

The goal of this study is to explain the influence of business coalitions. Influence is one of the key concepts in political science. While widely used in the study of nonstate actors in global politics, it oftentimes remains undefined and nontheoretical. As early as 1998, Michael Zürn (1998, 646) maintained that "although there is a lot of good evidence about the role of transnational networks in international governance, more rigid research strategies are needed to determine their influence more reliably and precisely." Since then, only a few scholars have answered this call (Arts 1998; Newell 2000; Betsill and Corell 2001, 2008). Building on this body of literature, and in particular on Michele Betsill and Elisabeth Corell's book, I define influence, propose indicators of influence, and suggest methodologies to identify evidence for influence in the following paragraphs.

Influence "occurs when one actor intentionally communicates to another so as to alter the latter's behavior from what would have occurred otherwise" (Corell and Betsill 2008, 24). While the concept of influence is intrinsically linked to power, the two differ from each other: power refers to the capabilities to affect political change, while influence refers to the actual effect of an actor on political outcomes. Different forms of power are the sources of influence, but power does not equal influence. Whether actor-related forms of power translate into influence depends on a number of variables, such as strategy and the political opportunity structure, as will be discussed in detail in the next two sections. In this respect, this study's influence-oriented approach differs from the great majority of analyses of the role of firms in global environmental politics, which focus mostly on power resources.

For now, the question is, what would be the appropriate indicators for nonstate actor influence, and how can we detect influence methodologically. Goal attainment, generally speaking, is the indicator for influence (Keck and Sikkink 1998). It is essential to focus on the goal attainment of nonstate actor political activity in the attempt to infer influence. Previous studies have often simply looked at the activities per se, access to decision makers, and the resource equipment of NGOs as a basis for suggesting nonstate actor influence (Betsill and Corell 2001). It is highly likely that these approaches have led to overestimating the influence of nonstate actors.

For goal attainment to be operational, it needs to be differentiated in more specific political goals: to "assess the influence of advocacy networks we must look at goal attainment at different levels" (Keck and Sikkink 1998, 25). Nonstate actors may pursue different goals in the course of the policy cycle. While affecting the political outcome is most likely the ultimate goal, agenda setting is also a major political goal of nonstate actors (Arts 1998; Corell and Betsill 2008). Especially in battles over meaning, agenda setting is the primary goal of nonstate actors because the agenda usually has significant impact on the political outcome (Sell and Prakash 2004). Hence, we acknowledge influence if nonstate actors have attained their political goals regarding the agenda and/or political outcome.

If nonstate actor influence is understood as the effect on the agenda and political outcomes, the remaining question is how we can infer causality between nonstate actor activities and the observed agendas and political outcomes. For a number of reasons, this is an intrinsically difficult task to achieve (Corell and Betsill 2008). Advocacy activities, for one, frequently occur in private, hidden behind a veil of secrecy. In interviews, decision makers may not want to disclose information on the influence of particular interest groups. This is most likely the case when issues are particularly controversial and/or lie in the immediate past. Second, NGOs often demand different things in public than in private, which makes public statements a potentially unreliable source.

Given these difficulties, I suggest a range of methods including best and second-best ways to infer a causal link between nonstate actor activity and agenda/political outcomes, including process tracing, correlation, and counterfactual analysis. The research goal of explanation warrants clarification to avoid confusion regarding the epistemological standpoint of this study. This book deviates from the neopositivist notion of explanation as establishing formal causality by building on the idea of narrative causality. The Newtonian idea of formal, or mechanical, causality assumes that single causes can be identified as the necessary and sufficient condition for an effect. Such neopositivist assumptions of causality continue to exist in the study of world affairs (see, for example, King, Keohane, and Verba 1994). Yet there is a growing awareness in the social sciences and international relations that "the social world is inherently indeterminate" (Ruggie 1995, 95). The causation of social phenomena is most often complex, and defies explanations based on a nontrivial necessary or sufficient variable. Multiple causes play a role in the emergence of social effects. In light of these assumptions, scholars have proposed an alternative understanding, which has been referred to as narrative causality (Ruggie 1995) or complex causality (George and Bennett 2004). Complex causality exists if "the outcome flows from the convergence of several conditions, independent variables, or causal chains" (George and Bennett 2004, 212).

Social scientists have increasingly relied on process tracing to establish narrative causality. Sometimes this method is also referred to as *thick description*. "The process-tracing method attempts to identify the intervening causal process—the causal chain and causal mechanisms—between an independent variable (or variables) and the outcome of the dependent variable" (George and Bennett 2004, 206). In the research process, process tracing requires two steps (Ruggie 1995). In a first step, events, or social facts, are identified and organized in chronological order, and their effects on each other are established to the extent possible. The second step is a dialectic process in which a story is woven by going back and forth between social facts and a theoretical theme. Such analytic process tracing organizes "descriptive statements into an inter-subjective gestalt or coherence structure. . . . The aim is to produce results that are believable and verisimilar to other observers of the same process" (ibid., 98). In other words, the goal is not to produce a law across time and space but rather to generate an analytic narrative that makes sense to others. Process tracing consequently is not conducive to developing generalizable and parsimonious theories (Checkel 2006). Instead, the approach produces causal narratives that are the "intermediate between laws and descriptions" (Elster 1998, 45). In the case at hand, process tracing would show how the political activities of a particular nonstate actor, such as providing information to decision makers, affect the political agenda and/or political outcome. While process tracing would be the method of choice, there are cases where a lack of evidence for the above-mentioned reasons prevents the use of this method.

A second-best method for such cases is correlation. If the agenda/political outcome reflects the preferences of a particular nonstate actor, this is most likely due to the influence of this actor (Corell and Betsill 2008). This method is not without flaws. Correlation usually relies on publicly available statements about nonstate actor preferences. As has been said, it is not granted that public preferences are the same as the policy preferences that interest groups communicate in private. Further, correlation can only be used to infer causality as long as no two actors in the game hold the same policy preference. Otherwise, the environment would no longer be controlled, rendering causal inferences impossible. Counterfactual analysis in these situations may be employed as an alternative second-best method. The guiding question, then, is whether the agenda/political outcome would be the same had it not been for the influence of a given nonstate actor. These three methods—process tracing, correlation, and counterfactual thought experiments—should be understood in the sense of an escalation ladder. If the evidence is insufficient for the use of method A, then method B will be employed, and so on. Two methods also may be used for the same case to triangulate findings.

Case Selection

The empirical part of this book is designed as a comparative case study. To explain the choice of cases, I take two steps. First, I justify the selection of the overall issue of emissions trading in climate politics as an area to study the role of business in global environmental governance. In a second step, I explain why I chose the three specific cases—that is, the Kyoto Protocol, the EU ETS, and U.S. trading schemes. I have selected the rise of emissions trading as an issue area for studying the influence of proregulatory business collective action for two reasons: it represents the single most significant example of the marketization of global environmental governance, and it is an empirically rich case.

For one, the spread of emissions trading represents the single most significant example of market-based global environmental policy in a number of ways. Global GHG emissions trading represents a strong form of marketization, as the regulatory instrument itself creates a new market. In this respect, it is well suited as a case to study the broader trend toward the marketization of global environmental politics. Moreover, the scale and scope of carbon trading are unparalleled compared to any other market-based policy initiatives. Emissions trading affects a great number of industry sectors in several key industrialized nations. Hence, it is a highly relevant case in terms of sectoral and global coverage. And finally, global climate change is considered the most pressing environmental problem of the twenty-first century and has come to define global environmental politics. The rise of market-based climate politics thus is also highly relevant with regard to the gravity of the environmental problem.

The second reason for the choice of the diffusion of emissions trading as a case is that it is an empirically rich example of corporate involvement in global environmental politics. Protrading coalitions emerged in a number of different political contexts. Such variance in variables allows for maximizing observations through across-case comparison.

This book compares three cases that align with historical periods and milestones in the globalization of carbon trading: the internationalization of emissions trading in the Kyoto Protocol (1995–2000), the emergence of the EU ETS (1998–2008), and carbon trading in the United States (2001–2008). These three instances have been selected because of their relevance to the overall process of the spread of carbon trading. Only the agreement of the Kyoto Protocol led to the internationalization of the U.S. regulatory instrument of emissions trading. The EU implemented the first large-scale trading scheme. The move of the United States toward emissions trading gets its significance from the importance of the United States in any effective international climate framework. In addition, the spread of project-based carbon trading to the developing world hinges on markets in the EU and the United States, as they represent the backbone of emerging carbon markets. The policy community widely perceives that the political action on carbon trading shifted from the international level to the EU and then to the United States. The case selection therefore is based on the chronological order of major events. Furthermore, the case selection allows for the analysis of business coalitions and their influence across political systems (the EU and the United States) and political levels (international, regional, and national).

Each case will explore the degree of business influence in the rise of carbon trading on the political agenda and the sources of this influence. Defining the boundaries of the unit of analysis of business collective action in each case study has proven to be difficult. When coalitions have a low degree of institutionalization, as in the run-up to the Kyoto Protocol and the case of the EU, it is often unclear who cooperates with whom. The reason is that corporate advocacy tends not to be transparent. The chosen solution to this problem has been to consider only those actors as part of the coalition whose membership could be confirmed by at least two independent sources—that is, through triangulation. Thus, the key actors could be identified, while actors at the margins might not be mentioned.

Data Collection

In order to collect relevant and comparable data, Alexander George and Andrew Bennett (2004) suggest the method of structured, focused comparison. The method is structured in that empirical research is guided by a set of research questions that are derived from the research objective. These questions are applied to all cases. The method is focused because it pays selective attention to particular analytically relevant aspects of historical cases.

Data come from both secondary and primary sources. In terms of the secondary sources, political scientists and management scholars alike have investigated corporate involvement in climate change politics, especially that of the oil sector. The broader empirically founded research on international climate politics has proven to be a valuable source as well. Next to academic research, secondary sources, such as reports (for example, by the Carbon Disclosure Project and the Coalition for Environmentally Responsible Economies CERES), industry journals (*Oil and Gas Journal* and *Environmental Finance*), policy reports (*Economist, ENDS Reports, Environment Reporter*, and *International Environment Reporter*), and newspapers/magazines (*Economist, Financial Times, New York Times, Wall Street Journal*, and *Washington Post*) have been critical sources. With regard to the political role of oil companies up to 2000, I was able to resort to an extensive collection of data based on documents and interviews compiled by Levy and Ans Kolk.[2] I obtained primary data from documents and semistructured interviews. The documents reviewed include primarily corporate reports, press releases, policy papers, policy reports, and notes from parliamentary hearings. Most of these documents were obtained through the Internet, although my interview partners also provided some of them. In addition, I conducted fifty-two elite interviews of an average length of one hour with managers (ten), representatives of business associations (fourteen), policymakers (twelve), NGO professionals (eleven), and other experts (five) in Europe and the United States. The majority of the interviews were conducted in London, Brussels, and Washington, DC, in July and August 2007. A second interview phase took place in Washington, DC, in January 2008. A number of phone interviews with individuals in Europe and the United States also were conducted throughout 2007 and 2008. For reasons of confidentiality, the list of interviewees could not be included in this book.

I selected my interview partners on the basis of how often they were referred to in articles and conference programs as well as by other actors in the field. The interview partners represented all three key sectors involved in climate change politics: business, government, and environmental groups. If the information could be triangulated through sources from at least two sectors, it was likely that the presentation of the social facts was not distorted by a particular group's vested interests. This was not always easy to determine, as many interview partners seemed to want to take credit for a particular political event. Therefore, the total number of interviews had to be increased to satisfy the triangulation requirements.

The interviews were designed to provide background in arriving at an intersubjective explanation for the rise of emissions trading and the emergence of protrading coalitions. They adhered to strict terms of confidentiality—that is, direct quotations were only allowed with the interviewee's written approval. In all other cases, the information provided has been given anonymity. Anonymity was guaranteed to the interview partners because of the sensitivity of corporate lobbying. This sensitivity stems from the fact that effective lobbying frequently requires nontransparency, as corporate lobbying has a strong competitive dimension and is tied to reputational risks. Moreover, these effects were aggravated by the fact that most of the events discussed are recent. Prior to each interview, the interviewee was sent an information sheet including the terms of confidentiality. These terms build on a template used for a human subjects' consent form provided by the London School of Economics' Methodology Institute. Each interview has a number by which it is cited in the book.

Overview of the Book

This chapter has so far introduced the case of the rise of carbon trading as a manifestation of the broader trend toward market-based environmental policy. I have argued that the emergence of a transnational, protrading business coalition has played an essential role in the global spread of emissions trading. While firms could not prevent carbon controls, they had critical influence in shaping the regulatory style. Alternative explanations as offered by neo-Gramscianism, neoliberal institutionalism, and constructivism suffer from theoretical deficiencies that render their readings of the diffusion of carbon trading unsatisfactory. They nonetheless offer a number of helpful vantage points for the analytic framework advanced in this study.

Chapter 2 presents a theoretical framework for studying the role and influence of business coalitions in global environmental politics. The central assertion is that coalitions provide firms with a power through organization, which allows them to leverage political resources and strategies. Regarding the former, the framework discusses the role of funding and legitimacy as factors determining influence. With regard to strategies, I discuss the ability to mobilize state allies and play political games at multiple levels. Beyond these actor-related factors, the chapter contends that political opportunity structures relate to how successful a transnational campaign is. Policy crises along with international and domestic norms particularly create opportunities or constraints for advocacy campaigns. I also introduce a two-by-two matrix of corporate political strategies in environmental politics, considering both the effects of institutions and compliance costs on strategy.

Chapter 3 considers emissions trading as a policy instrument, its intellectual history, and the state of the carbon market. Furthermore, the chapter explores how climate policy affects different industries and firms differently. To some industries, such as the oil, electricity, and energy-intensive manufacturing ones, engagement with climate policy is about managing and containing regulatory risk. Other industries including low-carbon technology producers, financial services providers, and investors can seize opportunities under a market-based climate regime.

Chapter 4 is the first original case study focusing on the internationalization of emissions trading by its inclusion in the Kyoto Protocol (figure 1.1).

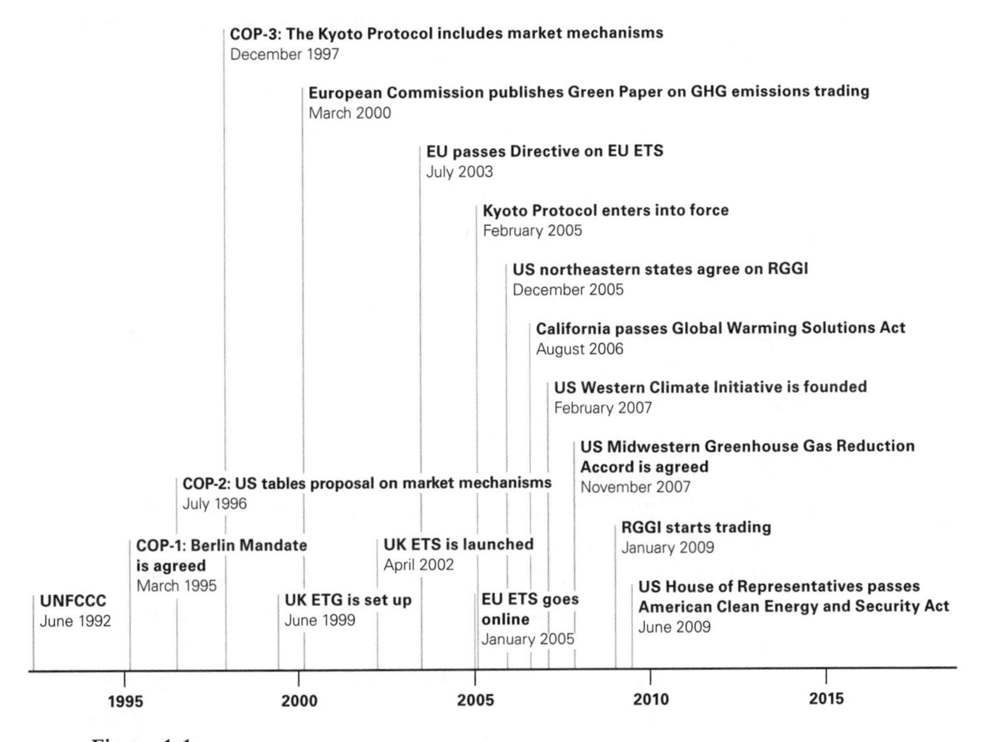

Figure 1.1
Milestones in the global rise of carbon trading

The chapter argues that a protrading business coalition emerged in the run-up to the Kyoto Protocol. This development represents a drastic shift from the early phase of climate politics in which business stood united in the opposition to emissions reduction targets. With BP and Environmental Defense at its core, the loose protrading coalition backed U.S. foreign policy, favoring a market-based climate regime in the international negotiations.[3] This policy led to the inclusion of emissions trading in the Kyoto Protocol. In the conference's aftermath, the transnational protrading lobby gained strength as new organizations such as the International Emissions Trading Association were created. Meanwhile, the antiregulatory business coalition had started to disintegrate.

Chapter 5 zooms in on European climate politics, exploring the bottom-up process of coalition formation in support of a Europewide emissions trading scheme. It shows how BP demonstrated the feasibility of a global trading scheme by implementing an in-house trading scheme. Subsequently, UK oil and electricity firms organized a domestic coalition for a pilot trading scheme in the United Kingdom. Corporate pioneers thus leveraged action at the European level. Oil majors BP and Shell as well as the European electric utilities found a powerful ally in the European Commission. Together these actors successfully introduced emissions trading to Europe, ultimately resulting in the creation of the EU ETS.

Chapter 6 analyses the rise of emissions trading on the agenda of U.S. climate politics and the emergence of a protrading NGO-business coalition. The chapter considers first the alignment of business and government strategies around voluntary climate policy under the Bush administration. It then explores how the bottom-up activities of business and U.S. states created political momentum for a domestic cap-and-trade scheme. This led, among other factors, to the emergence of the U.S. Climate Action Partnership, an influential protrading NGO-business coalition. The chapter looks at how the coalition shaped proposals for a federal climate bill.

Finally, chapter 7 summarizes the empirical and analytic findings on the role of firms in the rise of carbon trading. It also reconsiders potential alternative explanations in light of the empirical findings. The chapter concludes by discussing the power of business in the rise of market-based environmental governance, examining the effectiveness of market-based climate policy, and outlining the implications of this book for the study and practice of global environmental politics.

2

Business Coalitions in Global Environmental Politics

Business is an influential force in global environmental politics, not least because of its preponderance in economic resources. Firms have high stakes in the game, given that they are the primary entities regulated by environmental policy. Accordingly, they have a strong interest in influencing the rules of the game. They do so through lobbying, providing private governance, discursive activities, and technology investment, to name but a few dimensions of corporate political activities (Levy and Newell 2005; Falkner 2008; Fuchs 2008). The history of global environmental politics has taught us that the political behavior of firms has often turned out to be the key to success or the cause of failure in global environmental policymaking. One could in fact be led to believe that corporations rule the world of global environmental governance.

Yet the level of influence of firms in global environmental politics in particular and global politics more generally is not a foregone conclusion. While business occupies a privileged position, it also faces a number of constraints to its influence, as does any other interest group (Falkner 2008; Cerny 2010). First, countervailing forces, such as states and environmental groups, limit the power of business since they compete with business over influence (Wapner 1995). Their relative political weight and the level of contestation on a policy issue affect the influence of business. In addition, divisions in the business community cause conflict between and within business sectors, which constrains the overall political clout of business as such (Cox 1996b; Skidmore-Hess 1996).[1] The distributional effects of environmental policy and different institutional environments lead sectors and firms to different political strategies, which can result in inter- or intra-industry conflict.

Given these limitations to their influence, firms face the challenge of organizing collective action to achieve political clout, similar to any other interest group. As Cerny (2003, 156) notes, to quote him again, political outcomes "are determined not by simple coercion and/or structural power but, even more significantly, by how coalitions and networks are built in real-time conditions among a plurality of actors" (see also Mattli and Woods 2009). While business conflict limits the power of business, business cooperation enhances it. As mentioned in chapter 1, collective action generates power through organization. The power of collective action is indirect, as it helps to leverage other sources of power, including the pooling of material and immaterial resources, which are essential power resources (see Fuchs 2005a). The ability to organize collective action thus represents a strategic form of power (see Levy and Scully 2007). Despite the ubiquity of business collective action, the subject has received little attention in the study of business in global environmental politics. Yet transnational business coalitions, including their strategic allies among governments and NGOs, have often proven to be the political-economic engines driving or hindering progress in environmental politics. Against this backdrop, the key question is why some coalitions fail to influence policymaking while others succeed (see Newell 2008b).

In the following, I discuss transnational business coalitions as an increasingly prevalent form of the collective organization of firms in global environmental politics. I differentiate three types of coalitions based on their political strategy. I then continue to look at the role of political entrepreneurs in the creation and operation of alliances. Thereafter, the chapter proceeds by laying out a framework that explains the influence of transnational business coalitions in global environmental politics. I argue that business coalitions are strong and influential when contextual factors create opportunities for influence; political entrepreneurs are able to pool financial resources and acquire a high degree of legitimacy; and the coalition can mobilize powerful state allies and is capable of shaping policy at multiple political levels.

The argument reflects the understanding that both the political context of coalitions along with the characteristics and strategies of the coalition itself play into the success or failure of coalition advocacy. The international political context—in particular, policy crises and norms—creates opportunities and constraints for advocacy activities. Policy crises often open windows of opportunity that can be exploited by entrepreneurial coalitions that promote new policy solutions. Moreover, domestic and international norms create institutional opportunities or constraints for advocacy campaigns. The international context only creates opportunities and constraints for coalition advocacy; the coalitions' characteristics and strategies are crucial to their success or failure. Both the financial resources that a coalition pools and the legitimacy it acquires represent critical sources of influence. Successful coalitions also are capable of mobilizing powerful state allies and affecting political change across numerous political levels in multilevel environmental governance. These coalition characteristics and strategies are essentially a function of effective collective action, which reflects the strategic dimension of collective action.

The framework put forward here incorporates concepts from the study of business in global politics (see, for example, Fuchs 2008; Levy and Newell 2005; Falkner 2008) with concepts from social movement theory and transnationalism (see, for example, McAdam, McCarthy, and Zald 1996; Keck and Sikkink 1998; Sell and Prakash 2004; Betsill and Corell 2008). These research strands developed mostly separate from each other with little cross-referencing. The prominent concept of transnational advocacy networks, which in some ways resembles the transnational coalition concept, especially has been mostly applied to study the transnational collective action of NGOs. Yet as a number of scholars have argued, the study of firms and NGOs needs to be incorporated to advance theory making in the field of nonstate actors and transnational relations (Ruggie 2004; Ronit 2007). Drawing on both strands of research, the conceptual work that follows contributes to neopluralist perspectives on business in global environmental politics. It complements the analytic lens of business conflict with an analytic lens for business coalitions and their influence.

Business Coalitions, Strategy, and Political Entrepreneurs

Transnational coalitions of firms—often with NGOs or public actors—have become the political-economic power trains of global environmental politics either driving or stalling the process (see DeSombre 2000; Matthews and Paterson 2005). Nevertheless, these political animals remain understudied in international relations and international political economy. The literature on coalitions in social movements and transnational advocacy networks provides a rich source of concepts that can be brought to fruition in the study of transnational business coalitions (Heclo 1978; Salisbury et al. 1987; Sabatier 1988; Keck and Sikkink 1998; Meyer and Corrigall-Brown 2005). In the following subsections, I discuss why transnational coalitions have become an important form of interest aggregation for firms as opposed to sectoral trade associations. I then identify different types of business coalitions depending on their political strategy. Finally, I explore the crucial role of political entrepreneurs in forging alliances among firms and devising political strategy.

The Rise of Transnational Coalition Politics

Traditionally, business interest representation is organized in sectoral trade associations. In the last two decades, however, "complex multilevel and institutional advocacy coalitions" have increasingly evolved in Europe, the United States, and at the international level (Coen 1998, 2005). "Transnational coalitions are sets of actors linked across country boundaries who coordinate shared strategies or sets of tactics to publicly influence social change. The shared strategies or sets of tactics are identified as transnational campaigns" (Khagram, Riker, and Sikkink 2001a, 7). Transnational business coalitions are alliances of business actors and sometimes other nonstate actors as well as state actors brought together by the interest to push for or against a particular policy. I refer to them as business coalitions because firms are usually the key actors driving the coalition. Such coalitions emerged in virtually all the major environmental policy fields, including ozone, biosafety, and climate change (Levy and Newell 2000; Falkner 2008). But also beyond the environmental realm, firms created transnational coalitions, including in the intellectual property rights field (Sell 2003) and European market integration (Cowles 1995), to name but a few. A number of factors have contributed to this trend, such as the demand from public actors for the broad-based representation of interest groups, increasingly diverse interests among large firms, and the need to accommodate nonbusiness interests. "Thus, the most successful lobbyists are not necessarily those who paid the highest political contributions, but those who extract the broadest support from the greatest number of actors, and for this reason the US system, like the EU, can be seen to be based upon alliance building, bargaining, and compromises" (Coen 2005, 210).

Sidney Tarrow (2005) identifies four forms of coalitions that vary according to the intensity and duration of cooperation: instrumental coalitions, event coalitions, federations, and campaign coalitions. In instrumental coalitions, actors coalesce on a short-term basis with a relatively low level of involvement. Event coalitions also have a short lifetime, but involve stronger cooperation between actors around an international event. Federated coalitions are long term, but require a low level of commitment from their member organizations. Finally, campaign coalitions are long term and demand a high level of involvement.

Given the variation across the intensity and duration of cooperation, business coalitions show varying degrees of institutionalization (Tarrow 2005). While highly institutionalized coalitions are easy to detect, scholars often neglect noninstitutionalized ad hoc coalitions because they are more difficult to identify. Coalitions frequently are embedded in broader policy networks of actors that share interests and norms, but that do not necessarily coordinate their strategies (Mahoney 2007). A coalition-centered perspective on business involvement in global environmental politics allows analysts to identify broader lines of contention across time and space. Increasingly, conflict does not run between one industry sector and another, or between environmental and business groups. We instead find different alliances of fractions of business with public actors and/or environmental groups competing with each other across political levels (Newell 2008b).

Distributional Effects, Institutional Pressure, and Corporate Political Strategy

If interest aggregation does not simply occur along sectoral lines, along what lines do different business coalitions form? As the above definition says, actors in a coalition "coordinate shared strategies" (Khagram, Riker, and Sikkink 2001, 7). Against this backdrop, the analysis of corporate political preferences and strategies is a necessary step to understanding the lines along which business conflict and business coalitions emerge. Two main factors generally determine the formation of corporate political strategies: the economic effects of environmental policy on the firm, and the firm's institutional environment (Pinkse 2006; Falkner 2008).

Being distributional in nature, environmental policy affects different sectors and firms differently depending on their respective market position (seeCox 1996b; Skidmore-Hess 1996). Such distributional effects exist when environmental regulation causes lower aggregate costs to an industry as a whole compared to other industries; when environmental regulation generates rents for some industries or firms while it erects barriers to other industries and firms; and when environmental regulation causes differential costs across firms in the same industry (Keohane, Revesz, and Stavins 1998). As rational actors in a competitive environment, firms will assess the economic effects of environmental regulation—especially compared to their competitors. Some firms will win, and others will lose. While firms have economic interests, they are also embedded in wider social institutions that affect how they interpret their economic interests (DiMaggio and Powell 1991; Woll 2009). These institutions exist at different levels, such as the home country (Sally 1995; Doremus et al. 1998) or the level of organizational fields (Hoffman 2001). The interaction of the economic effects of environmental regulation on firms and the institutional effects influence the choice of corporate political strategy, and by implication, the lines along which conflict and coalitions emerge. I identify four types of

Table 2.1
Corporate political strategies in environmental politics

	Economic effects	
Institutional pressure	Costs > benefits	Costs < benefits
Low	Antiregulatory strategy	Nonparticipation
High	Proregulatory risk management strategy	Proregulatory market-making strategy

corporate political behavior in environmental politics, as shown in table 2.1: antiregulatory strategies, proregulatory risk management strategies, proregulatory market-making strategies, and nonparticipation. Below, I will explain under which circumstances firms adopt what strategy.

If an environmental regulation predominantly imposes costs on a firm, and if the institutional pressure for environmental action is low, the firm is likely to oppose the environmental regulation (antiregulatory strategy). As the public demand for environmental action increases, however—that is, the institutional pressure is high—an antiregulatory strategy bears considerable reputational costs. This will lead firms most likely to adopt a proregulatory strategy that advocates the least costly policy option to minimize compliance and reputational costs (proregulatory risk management strategy). Self-regulation has been widely used by firms as a low-cost regulatory option in a number of issue areas of regulatory politics (Haufler 2001; Falkner 2003; Prakash and Potoski 2006; Pinkse and Kolk 2009). It partially addresses the public demand for regulatory action, while mostly leaving firms in control of regulatory instrument design. Alternatively, if firms already operate under domestic unilateral environmental policy, they might advocate the internationalization of this policy to create a level playing field with their competitors (DeSombre 2000). In this way compliance costs are not minimized, but it is ensured that competitors are subject to the same costs. Such corporate behavior has been identified as a source of the "trading up" of environmental policy (Vogel 1995; Young 2003). Instead, if a firm stands to benefit from an environmental policy, it is unlikely to engage in proregulatory advocacy if the institutional momentum is low (nonparticipation). Political engagement is costly, and if the institutional pressure from governments and environmental groups to take environmental action is low, the chances are low that the policy will be implemented. Yet if the institutional pressure is high, firms that stand to benefit from environmental regulation are likely to advocate for the environmental regulation (proregulatory market-making strategy). This is because high institutional pressure increases the likelihood of environmental legislation being passed. The goal, then, is to expand existing markets or create new markets through environmental regulation (Levy and Prakash 2003).

While coalitions pursuing an antiregulatory strategy in most cases consist of firms only, proregulatory coalitions often include environmental organizations, as "firms have learned to *mix and match* their political alliances with various environmental and business interests groups to create flexible advocacy coalitions" (Coen 2005, 216). Coalitions of firms and green groups have been referred to as "baptist-and-bootlegger" coalitions (Yandle 1983; Vogel 1995; DeSombre 2000). The term connotes the cooperation of unlike interest groups in advocacy for a common cause.[2] Finally, a single coalition can also change its strategy from mobilizing against regulation to advocating a particular policy over the course of its lifetime, as it accommodates pressure from countervailing forces.

The phrase "perceptions of control" (Fligstein 1996, 2002) captures the policy goals around which business coalitions form. Neil Fligstein posits that firms generally aim for stability in the institutional and political underpinnings of markets. So-called organizational fields, the institutional environment of firms, are stabilized through conceptions of control, which are defined as "perceptions of how markets work that allow actors to interpret their world and act to control situations" (Fligstein 1996, 658). Basically, conceptions of control are norms that channel expectations in markets, thereby creating a stable institutional environment. In the case at hand, emissions trading can be seen as a conception of control in dealing with the climate change challenge as much as the outright rejection of emissions controls as a competing conception of control (see Pulver 2007).

Political Entrepreneurs and Coalitions

Shared political preferences are a necessary, but not sufficient condition for the formation and effectiveness of coalitions. The creation and operation of coalitions also rests on political entrepreneurs who believe that mobilizing other actors will further their mission (Keck and Sikkink 1998). The role of political entrepreneurs in organizing business representation is understudied. Social movement theory, though, offers a number of insights on the relevance and role of leaders in coalition formation and maintenance (Tarrow 1998; Barker, Johnson, and Lavalette 2001; Ganz 2009; Andrews et al. 2010). The key task of political entrepreneurs is "organizing," which "involves a host of particular skills, from persuading others of an idea or tactic to maintaining commitment and morale, from allocation resources to recognising opponents' weak points, from writing and speaking to taking initiatives in moments of decision, from forming (and breaking) alliances and conciliating opponents to issuing authoritative commands" (Barker, Johnson, and Lavalette 2001, 6).

As this definition reflects, the organizing role of political entrepreneurs has an internal and external dimension. Entrepreneurs mobilize other actors for an alliance, but they are also key in developing and devising the political strategy vis-à-vis third actors—notably governments. Actors that lead the coalition building process are so-called coalition brokers (Loomis 1986). They are at the coalition's core and usually have the most at stake, while firms in the periphery of a coalition are only marginally affected by the policy outcome (Hula 1999). Furthermore, coalition brokers tend to be large organizations, as those can muster the considerable resources necessary to successfully mobilize actors. These resources include money, human resources, and reputation. Mobilizing actors to join a coalition requires that the coalition broker probe "until he finds some new alternative, some new dimension that strikes a spark in the preferences of others" (Riker 1986, 64). In this process, actors might import innovative policy ideas from another country, thus acting as a vehicle of policy diffusion (Mintrom 1997).

Political entrepreneurs are also often the representatives of the coalition vis-à-vis government officials. They are therefore closely involved in formulating strategies and outreach activities. Frequently, the actors leading a coalition's political strategy are the initial coalition brokers. In transnational business coalitions, the political entrepreneurs and brokers typically are senior figures from public affairs departments, environmental departments, trade associations, lobbying firms, or environmental groups. Vocal CEOs sometimes assume the role of public figurehead of a coalition by giving public talks and meeting political leaders.

In the following, I will turn to the factors that play into the success of a transnational business coalition. Without delving into the agency-structure debate, I assume that it is an appropriate and pragmatic standpoint to consider both environmental factors and actor-related sources of coalition influence (Thacker 2000; Corell and Betsill 2008). The literature on social movements has come to recognize that both external and internal factors play into the efficacy or influence of social movement organizations (Schurman 2004). The literature on corporate power, in contrast, has largely focused on internal, capability-related factors to the neglect of structural and institutional determinants of influence (Fuchs 2005b; 2008). I also transfer insights from social movement theory to the study of business in global environmental politics. In line with these assumptions, I first discuss elements of the political context that offer opportunities or constraints for advocacy campaigns. I then explore coalition characteristics and strategies. While not explicitly included in the following framework as a factor of business influence, it is assumed that the level of interest group competition and the influence of countervailing forces affects the influence of a business coalition.

International Opportunities and Constraints

Contextual factors such as the institutional environment, power relations, and the political process affect the success or failure of advocacy (Corell and Betsill 2008). Social movement scholars refer to those factors as political opportunity structures (Useem and Zald 1982; Kitschelt 1986; McAdam, McCarthy, and Zald 1996; Campbell 2005; Tarrow 2005). A political opportunity structure is "a set of formal and informal political conditions that encourage, discourage, channel, and otherwise affect movement activity" (Campbell 2005, 44). As the definition reflects, the structure of the political environment can either provide opportunities for advocacy or constraints. In global environmental politics, such opportunities and constraints are "multilayered," as domestic and transnational political contextual factors interact (see Keck and Sikkink 1998; see alsoRisse, Ropp, and Sikkink 1999; Khagram, Riker, and Sikkink 2001a). A great many contextual factors affect advocacy success such as the rules of access, the stage in the policy cycle, the political stakes of the issue, international institutions, the level of contention, and the relative power of political adversaries (Khagram, Riker, and Sikkink 2001b; Tarrow 2005; Betsill 2008b). For the sake of parsimony, I focus on two contextual factors that play pivotal roles in global environmental politics: policy crises, and norms.

Policy Crises

Major policy changes are often preceded by exogenous shocks or policy crises (Haggard 1990; Sabatier and Jenkins-Smith 1993; Sell and Prakash 2004). As a policy crisis usually leads policymakers to question conventional policy wisdom, it opens a window of opportunity for new policy ideas. Policy entrepreneurs can capitalize on the opportunity by framing the policy issue in a new way and proposing a solution (Baumgartner and Jones 1993). Thus, a crisis can catalyze a shift in the balance of power from the incumbent coalition to one challenging the status quo.

The nature of such policy crises varies considerably. According to John Kingdon (1995), a crisis can occur due to changes in either the problem stream or the political stream. Translated to the field of environmental politics, crises in the problem stream include scientific discoveries about environmental problems and environmental catastrophes. They question the conventional wisdom about a particular environmental problem. Hurricane Katrina, for example, created a policy window for U.S. climate change politics by demonstrating the severity of the issue and the domestic impact of global climate change. In fact, many environmental activists consider crises in the problem stream, notably environmental catastrophes, as an essential precondition for major environmental legislation. Crises in the political stream of global environmental politics instead include the failure of existing policies, a political stalemate between adversarial groups, or the election of a new political leader. For instance, the stalemate between environmental interests, on the one hand, and business interests, on the other hand, in early climate politics presented a classic policy crisis. Policy crises or exogenous shocks, in short, create opportunities for new coalitions and advocacy campaigns. These opportunities relate primarily to the political process and power constellations, whereas the normative environment in which firms operate create other opportunities or constraints.

Domestic and International Norms

Domestic and international norms are crucial elements of the political environment of business coalitions. Norms not only shape the interests of firms (Woll 2009) but also enable or constrain their advocacy activities. Inspired by the new institutionalism in sociology, constructivist scholars introduced the "social fitness" notion to international relations (Checkel 1993; Weber 1994). The concept is that some ideas have a better fit with the normative environment than others. Accordingly, I hypothesize that the better the fit of a coalition's policy idea with the normative context, the more likely is its success. Norms exist at both the domestic and international levels.

First, different policy ideas resonate to different degrees with the domestic norms as embedded in ideologies and policies (Risse, Ropp, and Sikkink 1999). This is true for environmental politics, where scholars have identified different national styles of regulation (Vogel 1986). Drawing on the literature on varieties of capitalism, I distinguish two sets of domestic norms and institutions that matter for the adoption of environmental policy instruments: liberal market economies, and coordinated market economies (Wilson 2003; Hall and Soskice 2004; Mikler 2007). In liberal market economies, the primary coordination mechanisms are hierarchy and competition, which bring a set of institutions and norms with them. In coordinated market economies, firms also rely on nonmarket relationships in networks and other forms of collaboration. Firms in liberal market economies are used to react to clearly specified regulations, particularly to those regulations that work through the price mechanism. Market-based instruments, such as taxes and emissions trading, are the key regulatory tools in liberal market economies. Firms in coordinated market economies will react most efficiently to negotiated and agreed-on rules, including forms of industry self-regulation and voluntary approaches. Decision making between government and business is essentially consensual. Also, trade associations in coordinated market economies hold the authority to impose commitments on their members. Generally speaking, liberal market economies have an institutional and normative preference for market-based environmental policy, whereas coordinated market economies favor negotiated agreements. Hence, market-based policy ideas show a greater fit with liberal rather than coordinated market economies.

Second, norms can also be institutionalized at the international level, thereby creating international political opportunities or constraints for transnational advocacy campaigns. Constructivists have frequently pointed to the existence of internationally embedded norms that shape the interests and identities of state and nonstate actors (Adler 1997). For the environmental realm, Bernstein (2001) argues that underlying liberal norms have guided international environmental policymaking, which gave rise to what he refers to as "liberal environmentalism." Against this backdrop, coalitions that promote ideas commensurable with norms embedded in the international system have greater chances of success with their advocacy campaign. In a similar vein, international norms represent opportunities for advocacy since they represent a resource for coalitions to pressure governments into complying with the norm (Finnemore and Sikkink 1998; Keck and Sikkink 1998; Price 1998; Bernstein and Cashore 2000). Such international standards of appropriate behavior in global environmental politics relate to whether action should be taken or what kind of action is considered appropriate.

The coexistence of domestic and international norms raises the question of norm hierarchy: Do norms at one level trump norms at another level? While some scholars answer this question in favor of the primacy of domestic norms (Pauly and Reich 1997), I contend that the answer is subject to empirical research in each given case. The interaction of domestic and international norms creates multilayered opportunities and constraints for advocacy. For instance, Keck and Sikkink (1998, 73) assert that international norms sometimes can override domestic norms, if they are framed to connect to "domestic concerns, culture, and ideology." The relative relevance of domestic and international norms therefore needs to be investigated empirically. In sum, policy crises along with domestic and international norms represent contextual factors that promote or inhibit the success of particular coalitions. They represent necessary but not sufficient conditions for advocacy success. To what extent opportunities are effectively exploited by firms depends on their agency. This leads me to consider which power resources and strategies are at the disposal of business coalitions. I identify two critical resources—funding and legitimacy—and two main strategies—the ability to mobilize state allies, and play multilevel games.

Coalition Resources: Financial Resources and Legitimacy

Firms hold leverage over a number of resources, and different resources provide firms with different forms of power. The concept of power refers to capabilities owned by actors (Corell and Betsill 2008). Power is a crucial input factor for influence, but not tantamount to it. As this book argues, influence emerges from the interaction of environmental factors with actor resources and strategies. A number of scholars working on corporate power in global environmental politics have put forward a multifaceted approach to business power, including instrumental, structural, and discursive power (Levy and Egan 1998; Fuchs 2005a, 2008; Falkner 2008; Clapp and Fuchs 2009a). In this framework, I focus on two sources of power for reasons of parsimony: financial resources, and legitimacy. Funding is a source of instrumental power, whereas legitimacy is a pivotal source of discursive power (Levy and Egan 1998; Fuchs 2005a). From a pluralist perspective, the structural power of business factors implicitly into these forms of power. The larger the economic output or market share of a firm—forms of structural power—the larger, usually, are the funds available for lobbying and—by extension—the instrumental power of the firm. Similarly, if a firm contributes a large share to economic output or has a large market share, this may enhance its legitimacy vis-à-vis policymakers, which is a form of discursive power. It is important to note that the strategic value of collective action with regard to power lies in leveraging and pooling resources in the competition over influence (McCarthy and Zald 1978; Berry 1989; Hula 1999).

Pooling Financial Resources

Financial resources—and the organizational, technical, and human resources that come with it—matter in lobbying (Baumgartner and Leeth 1998; Fuchs 2005b; Coen 2007). Material resources are a source of instrumental power, which refers to "relations between actors that allow one to shape directly the circumstances of another" (Barnett and Duvall 2005, 49). In other words, instrumental power is a form of power that grants actor A control over actor B (see Dahl 1957). The fungibility and range of material resources play a role in at least two ways: to fund party contributions and lobbying activities, and run public campaigns.

First, financial resources allow firms to contribute to politicians' election campaigns, which may buy those firms favors in the future (Fuchs 2005b). Especially in the pluralist political system of the United States, campaign financing plays an important role in business-government relations. In the 2000 elections, firms spent $1.2 billion on campaigns, which is about fourteen times the amount spent by labor unions and sixteen times the contributions of other interest groups (United Nations Development Programme 2002, 68). Firms mostly contribute to campaign and party financing in their home countries. Financial resources, moreover, are essential to funding the organizational and human resources to engage with long, protracted policy processes. Oftentimes, lobbying success depends on long-standing relations and networks between lobbyists and policymakers (Levy and Egan 1998). Such networks are a function of a number of factors, including the availability of financial resources for maintaining a permanent office with skilled lobbyists. At the international level, policy processes move slowly and have multiple intervention points, which makes it particularly costly to engage with these processes. The availability of funds not only affects the sustainability of political representation but also the total number of business interest groups compared to other interest groups. It is estimated that a little more than half of all interest groups in the EU are business groups, while only one-third are public interest groups (Greenwood 2007, 13).

Second, funding is required for the staff, consultants, events, publications, and advertisements necessary to run effective advocacy campaigns. These inputs represent the material underpinning of "strategic social construction." Strategic social construction is a process "in which actors strategize rationally to reconfigure preferences, identities, or social context"

(Finnemore and Sikkink 1998, 888). It generates discursive power, which rests on "symbols, story-lines, and the provision of *effective* evidence and compelling arguments in the public debate" (Fuchs 2005a, 777). For instance, in the attempt to prevent agreement on an international climate treaty at the Kyoto conference, the Global Climate Coalition spent $13 million through the Global Climate Information Project, which stressed the economic implications of a climate treaty, notably for fuel prices and employment (Levy and Egan 2003). Money was spent on funding the project's organization and staff, commissioning reports, and running advertisements in newspapers and magazines as well as on television. Hence, financial resources perform an important part in generating discursive power (Fuchs 2005b). This connection between material and ideational sources of power is noteworthy. Especially in the global environmental politics field, ideas and notions of legitimate behavior are frequently the primary currency.

Given the high cost of political activities, it is clear that financial resources matter. Yet I want to caution against assuming a linear relationship between financial resources and political influence. This is particularly true when financial resources are used to influence political discourses, as discourses are not controlled by single actors but rather emerge organically from the cacophony of voices in public debates (Levy and Egan 1998). As I have previously argued, influence depends on a number of actor-related and contextual variables. Immaterial resources such as legitimacy or expertise, furthermore, might compensate for a lack of financial resources. As John Heinz and his colleagues (1993) maintain, the effect of differences in resource equipment among interest groups on their relative influence varies across policy issues. This ultimately presents an empirical question warranting contextual analysis.

Acquiring Legitimacy through Cooperation with NGOs

If a business coalition has a high level of legitimacy in the eyes of policymakers, chances are higher that its advocacy efforts will be successful than with a low degree of legitimacy. Legitimacy is a contested concept in political science (see Clark 2003). For the purpose of this study, I adopt the following definition: "Legitimate behaviour is rightful behaviour: undertaken by the appropriate authority, in line with an agreed set of rules, and with appropriate or intended effects" (Collingwood 2006, 444). The idea that an actor is legitimate implies that there is normative consent on the actor's part granting the legitimacy (Hall and Biersteker 2002). Usually, legitimacy is the product of persuasion, social capital, or generally noncoercive forms of interaction. It is the vital resource to turn material power into authority (Cutler, Haufler, and Porter 1999). Raw instrumental power alone is often insufficient to convince policymakers of an idea. As David Coen (2007, 335) argues for the case of the EU, lobbying is a "credibility game" that requires long-term, trust-based relationships between representatives of interest groups and government officials.

Legitimacy is a political resource that generates productive, or discursive, power. "In order to effectively exercise discursive power in the political process an actor requires legitimacy as a political actor. After all, discursive power is relational in that it relies on the willingness of recipients of messages to listen and to place at least some trust in the validity of the contents of the message" (Fuchs 2005a, 779). There is evidence to assume that firms often have a lower level of legitimacy than do environmental NGOs. The disembedding of global capitalism through neoliberal economic policies in the 1990s has spurred attacks of business by the antiglobalization and global environmental movements. This in turn has eroded public trust in business, resulting in a lack of legitimacy of firms in the public eye. The corporate social responsibility movement can be interpreted as exactly such an effort to tackle the legitimacy deficit as corporations are adopting environmental and social standards to retain their moral license to operate (Ougaard 2006; Ruggie 2007).

While there are several paths for nonstate actors to being recognized as a legitimate voice including expertise (Haas and Adler 1992), I argue that business coalitions in global environmental politics rely in particular on cooperation with environmental groups as a source of legitimacy. Firms acquire legitimacy by cooperating with actors that possess complementary resources. It is often assumed that the key asset of NGOs is legitimacy, which gives them a comparative advantage vis-à-vis firms when it comes to discursive power (Simmons 1998; Raustiala 2002).

NGO-business coalitions as well as partnerships have emerged since the mid-1990s as an essential element of the strategies of corporations and business-oriented, cooperative environmental organizations (Bendell 2000; Pattberg 2007). The literature on NGO-business partnerships has frequently pointed out that one of the motivations of firms to collaborate with NGOs is to legitimize their corporate conduct (Murphy and Bendell 1997). As mentioned earlier, NGO-business coalitions that pursue a pro-regulatory strategy have been referred to as baptist-and-bootlegger coalitions, because two groups that are unlikely to cooperate find themselves working for the same goal (Yandle 1983). In global environmental politics, firms are the bootleggers, while green groups represent the baptists

(Vogel 1995; DeSombre 2000). While environmental groups bring legitimacy to such coalitions, their high level of legitimacy also stems from the fact that antagonistic interest groups advocate a compromise solution. Policymakers look for signs that a policy idea has broad support across interest groups (Kingdon 1995). Baptist-and-bootlegger coalitions supply just that. They not only signal broad support to policymakers but also signal that differences between two adversarial interest groups have been resolved (Hula 1999).

NGO-business cooperation is used as part of not only proregulatory but also antiregulatory strategies. The mobilization of civil society groups by firms to create broad opposition to regulation is known as "astroturf organizing" (Stauber and Rampton 1995). The idea is to enhance a cause's legitimacy by creating the impression that there is a grassroots movement in support of it. For instance, the Global Climate Coalition set up the front group Global Climate Information Project, which mobilized allies such as labor unions and farmers to oppose an international climate treaty (Levy and Egan 2003). To conclude, coalitions allow firms to pool crucial political resources, notably funds and legitimacy. These capabilities are the sources of instrumental and discursive power. Yet they only translate to influence if they are skillfully employed.

Coalition Strategies: State Allies and Multilevel Games

Influence is a function of proficiently employing resources under given circumstances. I have discussed key contextual factors—policy crises and norms—as well as resources—funding and legitimacy. Now I turn to strategy as a source of influence. A burgeoning literature has made the point that strategy matters for the success of advocacy campaigns (Oliver 1991; Fligstein 1997; Getz 1997; Lawrence 1999; O'Brien et al. 2003; Hillman, Keim, and Schuler 2004). Strategy refers to "the targeting, timing, and tactics" through which resources are mobilized and deployed (Ganz 2000, 1005). Strategy is closely linked to the role of political entrepreneurs in transnational business coalitions. The effectiveness of strategy depends on an organization's strategic capacity, which in turn is partly a function of political leadership (ibid.). The focus here, however, is on identifying particularly effective strategies. Coalition building is, of course, a key political strategy for firms—the basic argument of this study. Yet collective action allows firms to leverage other strategies. Two of these have received little scholarly attention, although they appear to have proven effective:

the mobilization of state allies, and the strategy of targeting multiple levels in global environmental governance to affect broad-based policy change.

Mobilizing State Allies

In the complex multilateralism of global governance, nation-states remain the primary actors (O'Brien 2003; Drezner 2007). They represent the "'coral reefs' in a broader sea of complex internationalism" (Tarrow 2005, 27). Given the central role of the nation-state in global environmental governance, business and other interest groups rely on state allies to exert political influence (Thacker 2000; Betsill 2008b). This statement might seem tautological: if firms have a state ally, one might argue that they will no longer have to persuade government actors of their policy preference. This would be true if the state were a monolithic actor. Different state agencies, however, often have different policy preferences, as the state is a fragmented, disaggregated entity (Slaughter 2004). This creates opportunities for coalitions to seek an alliance with a particular agency in the attempt to outcompete other agencies in agenda setting and policy formulation. Moreover, in international politics a business coalition may team up with a few states in the pursuit of convincing other states of a specific idea. In short, because policy preferences differ across and within governments, the state is both an ally and a target. It will be made clear in the empirical analysis which state actors are allies and which ones are the targets of advocacy.

As studies on business-government relations have shown, firms have more influence at the domestic level of their home country compared to international negotiations (Levy and Egan 1998). This suggests that key political entrepreneurs in a coalition will seek government allies from their home country. The idea that business and state actors create alliances contradicts the general view that business and the state compete against each other in global environmental politics (see Betsill 2008b). The interests of firms and their home governments may converge on a number of issues. The "competition state" (Cerny 1990) is a likely ally for business coalitions that aim to prevent unilateral environmental regulation. It is frequently in the nation-state's interest not to burden its national economy with higher regulatory costs than those of competing economies. Business and state interests can also align in support of environmental legislation. For instance, in the case of existing unilateral environmental policy, firms and their home governments may push for the international adoption of their domestic policy in order to create a level playing field (DeSombre

2000). Or firms and governments may seek the international harmonization of different national environmental standards to eliminate trade barriers (Vogel 1995). While there are many cases in which state and corporate interests align, convergence is by far not always the case. The regulatory state often also acts as an adversary of business coalitions. In such cases, firms might try to exploit intrastate conflict by aligning with a faction of government or seek state allies at the international level.

The mobilization of state allies is a function of both the strategies of political entrepreneurs and the activities of government actors. Political entrepreneurs from business build good relations and trust with policymakers, thus establishing networks from which coalitions can emerge. Firms also organize broad coalitions to increase the legitimacy of their cause vis-à-vis policymakers. Hence, there is considerable agency in the pursuit of powerful state allies. Yet the availability of state allies also depends on the preferences and activities of government actors. State actors themselves sometimes actively intervene in the organization of interest representation to manage the input they receive from lobbying groups (Woll 2009). This can serve different purposes such as generating support for a particular policy or simply reducing the transaction cost of state-business interaction. For example, the European Commission created policymaking forums and select committees in response to a boom in public interest lobbying (Coen 2005). This, in turn, created incentives for coalition formation among business actors. As such, when studying state allies, close attention needs to be paid to how the two sets of actors interact and where the agency lies. Next to mobilizing state allies, a second strategy has proven critical to the success of transnational business coalitions: the ability to effect change across political levels.

Playing Multilevel Games

Robert Putnam (1988) famously observed that domestic and international politics are closely intertwined in two-level games. In many cases, global governance has become a game of even higher complexity in which multiple levels of policymaking interact (Hooghe and Marks 2003). Global agendas and political outcomes are determined by processes as well as events at the subfederal, national, regional, and international levels, as political authority is highly decentralized in systems of multilevel governance. Large-scale global policy shifts thus require change at a number of political levels. Given this, the success of business coalitions also hinges on how well they are capable of affecting political change at multiple political levels in the complex system of global governance. In the negotiations of the Agreement on Trade-Related Intellectual Property Rights, for instance, a transnational business coalition could successfully graft its policy preference on the international agenda because, among other reasons, it simultaneously lobbied the governments of the United States, the EU, and Japan—crucial powers in the international process (Sell and Prakash 2004).

Playing multilevel games requires in most cases the capacities of a transnational coalition. Individual firms can rarely effect such change. Firms often hold strong political capital in their home country, but are weaker at other political levels (Levy and Egan 1998). Moreover, engaging with multiple political processes at the domestic and international levels as well as in third countries generates costs that are generally prohibitive for even multinational corporations. Transnational coalitions instead are *plugged into* a range of political processes. Usually, member firms have public affairs offices in their home governments or are members in domestic trade associations. The coalition benefits from the political clout that its members have in different countries, and the local knowledge and networks with government officials. Furthermore, coalitions allow firms to pool sufficient financial, organizational, and human resources to engage with a number of strategically important policy processes. To leverage these advantages of a transnational coalition, firms need to learn to coordinate their activities with regard to message and timing. While firms by and large possess political capacities at the domestic level, they frequently need to develop these at the international and transnational levels.

Multilevel strategies play an especially significant role in setting global agendas for particular policy solutions. Global agendas emerge from political discourse across levels and are shared by many actors. Agenda shifts therefore require discourse to be influenced at a number of key levels. Influencing global political discourse through multilevel strategies is closely tied to the strategy of framing. As Sanjeev Khagram and his colleagues (2001a) argue, framing is the primary strategy in battles over meaning. Through framing actors influence policy debates, set agendas, and ultimately affect political outcomes (Baumgartner and Jones 1993; McAdam, McCarthy, and Zald 1996). Frames are "specific metaphors, symbolic representations and cognitive clues used to render or cast behaviour and events in an evaluative mode and to suggest alternative modes of action" (Zald 1996, 262). To be clear, frames are not ideas but rather ways of packaging and presenting ideas and information. The notion of framing has been developed in the social movement literature (Snow and Benford 1988), but it resonates with the idea of strategic social construction in the constructivist research strand in international relations.

Through their evaluative mode, frames help to make sense of information. Given bounded rationality, actors rely on the filter and evaluative functions of frames to deal with the wealth of information they are exposed to (Jones 2001). This is particularly true for the case of global environmental politics, which is a highly information-intensive issue area. Scientific information about the state of the natural environment plays a role next to information about the economic effects of environmental policy. Effective framing portrays change projects in such a way that they connect to the interests of affected actors. The better a change project links to the interests of others and underlying norms, the greater is its "frame resonance" (Snow and Benford 1988), or in other words, its social fit. While there is a strong structural component to fit—as discussed with regard to political opportunities created through norms, the idea of framing suggests that actors have some leeway in packaging ideas to improve their fit with external structures. Moreover, agency lies in the opportunity of "venue shopping" (Baumgartner and Jones 1991; Schneider and Teske 1992), which often goes hand in hand with framing. Venue shopping means that advocacy efforts are targeted at the most receptive venue. In a multilevel strategy, firms may not target several levels at the same time but instead focus their activities on receptive key levels of political action.

To conclude, since norms represent a form of social or institutional structure, no single actor holds control over them. Change projects therefore require a number of actors to simultaneously change norms across actors and political levels. Without multiple advocates promoting a particular policy idea across levels of political discourse and decision making, it is unlikely to be institutionalized. Next to external factors and political resources, political strategy is a factor playing into the efficacy of advocacy organizations. For transnational business coalitions, mobilizing state allies and affecting change at multiple political levels have been identified as essential strategies.

Summary: An Analytic Framework

Business has emerged as an influential force in global environmental politics. While firms have proven to be influential, they do not determine the outcome of global environmental politics. The level of firms' influence varies, to a large extent, with the level of conflict versus cooperation among firms. While scholars of global environmental politics have emphasized the possibility of conflict among factions of business as a factor limiting the power of business, little stress has been put on the role of collective action as a source of strategic power—that is, power through organization. Yet firms, similar to many other interest groups, rely on collective action as a major strategy in interest group competition. This is a main tenet of neopluralism. Traditionally, interests are aggregated through trade associations along sectoral lines. Increasingly, however, firms organize in transnational cross-sectoral coalitions in global environmental politics. These coalitions along with their NGO and state allies have become the political-economic engines of global environmental politics, affecting success or failure. Such a coalition perspective allows us to capture broader patterns of cooperation and contention in global environmental politics.

In global environmental politics, firms form different kinds of coalitions under different circumstances. The economic effects of environmental regulation on firms and the institutional environment are two critical factors affecting along which lines coalitions emerge. I differentiate three ideal types of business coalitions: antiregulatory coalitions—they oppose environmental regulation; proregulatory risk management coalitions—they advocate the least costly policy option; and proregulatory market-making coalitions—they promote policies that create primary and secondary markets. The emergence of transnational business coalitions depends on not only economic effects and institutional pressure but also the role of political entrepreneurs. These are often senior figures from business, trade associations, lobbying firms, or environmental groups who play prominent roles in brokering coalitions and devising political strategy.

The influence of transnational business coalitions is a function of how skillfully actors employ their resources under given opportunities and constraints. By combining contextual factors with coalition characteristics and strategies, this chapter puts forward a framework for studying the influence of business coalitions. Political opportunities or constraints emanate from policy crises as well as international and domestic norms. A policy crisis frequently leads policymakers to question conventional policy wisdom. This opens a window of opportunity for nonstate actors to propose new policy ideas. Furthermore, policy ideas show different degrees of fit with norms embedded in domestic political systems or the international level. The better the fit, the more likely governments will adopt the idea. Some environmental policy instruments show a better fit with liberal market economies in the Anglo-Saxon world than with coordinated market economies in continental Europe, and vice versa.

The organizational power of coalitions weighs in when it comes to shared resources and coordinated strategies. The key resources of coalitions are financial resources—and the organizational and human resources that come with it—and legitimacy. The fungibility of financial resources matters with regard to party contributions, maintaining a public affairs office, and being able to run public campaigns. A common and effective way for firms to increase the legitimacy of their causes in global environmental politics is the cooperation with green groups. In terms of coalition strategies, I argue that mobilizing state allies and playing multilevel games are both crucial. If the policy idea of a coalition is supported by a powerful state, it is more likely that the idea will affect political outcomes. As policymakers look out for signs of broad support, cross-sectoral coalitions are essential in mobilizing state allies. Finally, the success of coalitions also depends on their ability to drive change at several levels in the complex system of global environmental governance. With political authority dispersed across subnational, national, regional, and international levels, global change requires advocacy at a number of these levels. Transnational coalitions have a distinct comparative advantage here.

The framework presented here combines contextual factors with the power resources of coalitions and coalition strategies, going beyond a purely power-oriented—that is, capability-oriented—perspective on the role of business in global environmental politics. It claims to hold considerable explanatory power for variation in coalition success, yet it does not claim to be comprehensive in terms of covering factors relevant to the influence of nonstate actor coalitions. The case study will reveal to what extent they actually represent sources of influence. The conceptual work laid out in this chapter is understood as an extension of neopluralist approaches to the role of business in global environmental politics. It complements the study of conflict among business groups with a concept of business coalitions in global politics. After all, conflict and coalitions are the two sides of the coin of pluralist political competition. I now proceed with the empirical part of this book. The following chapter introduces the phenomenon of global carbon trading, and analyses the corporate interests and strategies in global climate politics to understand along which lines competing carbon coalitions emerged. Thereafter, I will tell the story of the global rise of carbon trading, applying the framework developed in this chapter. The focus is on the evolution of the coalitional politics of business and the sources of influence that the coalitions draw on.

3

The Political Economy of Carbon Trading

"The ideas of economists and political philosophers, both when they are right and when they are wrong, are more powerful than is commonly understood," John Maynard Keynes (1936, 383) once famously observed. Emissions trading is indeed an economic idea that has had tremendous influence on how governments, international organizations, companies, and consumers respond to the challenge of global climate change. Though nascent and fragmented, a global carbon market has been emerging over the last decade, with the prospect of significant growth in geographic scope and financial scale over the coming decade. This chapter draws the contours of the phenomenon of emissions trading in general and carbon trading in particular in four steps.

First, emissions trading is defined as a policy tool, whose unique features are emphasized by means of comparison with pollution taxes. Different forms of emissions trading are also explained. The intellectual and policy history of emissions trading is then portrayed, covering the period from its theoretical beginnings in the 1960s to its diffusion in the 1990s. In a third step, the geographic scope and financial scale of the current global carbon market are outlined. Finally, the link between business and emissions trading is established by discussing the distributional effects of climate policy and emissions trading on different business sectors and individual companies. This will result in a topography of business interests in climate politics, setting the stage for the analysis of business political action in the following case studies.

Explaining Emissions Trading

Emissions trading is a market-based environmental policy instrument, which is unique in that it creates new commodity markets. The instrument has outrun alternative policy instruments for climate policy, including pollution taxes, regulatory standards, and voluntary agreements. This is not to say that these policy instruments do not matter; rather, emissions trading has become the central pillar of climate policy mixes in the industrialized world. As this book argues, the rise of carbon trading can partially be understood as a strategy of big emitters to prevent the introduction of carbon taxes. Given this political-economic dynamic, in the following I will discuss the features of emissions trading in direct comparison to pollution taxes. I then turn to the two main types of emissions trading schemes: baseline-and-credit systems, and cap-and-trade ones.

Emissions Trading versus Pollution Taxes

Since the early 1970s, market-based environmental policies have found widespread adoption in the United States and other industrialized countries, complementing conventional command-and-control regulation. Among market-based environmental policy instruments, emissions trading and pollution taxes have emerged as the two options for mandatory climate policy. These instruments are meant to create incentives for environmental protection by putting a price on pollution. "Emissions trading . . . refers to the use of transferable rights, allowances, or credits in programs to control emissions" (Gorman and Solomon 2002, 293), whereas pollution taxes internalize economic externalities by charging polluters a fee or tax. Certainly in the United States, but also increasingly in other industrialized countries, emissions trading has been trumping pollution taxes in environmental policymaking (Stavins 1998). Yet the debate over the best approach has been going on since emissions trading was first proposed in the 1960s.[1] The characteristics of emissions trading as compared to pollution taxes are best demonstrated by the criteria of environmental effectiveness, economic efficiency, and distributional effects/political feasibility.[2]

Environmentalists have come to support emissions trading because it provides quantity certainty, meaning that the overall emissions reduction cap ensures a certain amount of emissions reductions (Chameides and Oppenheimer 2007). Quantity certainty comes at the cost of price volatility, however, which can deter the adoption of low-carbon technologies, as the long-term economic viability of these technologies is uncertain if future permit prices are unknown (Parry and Pizer 2007). Pollution taxes instead guarantee price certainty, which provides investors with planning certainty, but does not guarantee the attainment of a particular environmental goal. Some environmental groups therefore question the environmental effectiveness of taxes. As for economic efficiency, emissions trading and pollution taxes are on a par, economists say (Stavins 2007). Interestingly enough, in the political process emissions trading has always been portrayed as being more cost-effective than carbon taxes. Controversial issues regarding the expense of the two instruments relate to the transaction costs of carbon trading and the so-called double burden of pollution taxes, which arises because firms have to pay both abatement costs and taxes.

In terms of environmental effectiveness and economic efficiency, the academic debate has not resulted in a clear judgment on the superiority of one instrument over the other, although the majority of economists favor pollution taxes due to their simplicity and the price certainty they provide. While the former reduces options for rent seeking as compared to tradable permit schemes, the latter stimulates investment in pollution-reducing technologies. Nonetheless, the political process has preferred emissions trading. The real reasons for why emissions trading has been trumping carbon taxes lie in the political economy and distributional effects of those two instruments. Emissions trading serves the interests of many actors involved (Stavins 1998).

Business tends to favor emissions trading over pollution taxes because practice has shown that the majority of tradable permits are allocated free of charge, which is referred to as "grandfathering." Legislators thus buy industry acceptance of mandatory emissions controls, while business is given opportunities for rent seeking in the allocation process. These are not just one-off rents but instead can be sustained, as new entrants to the market need to buy permits from incumbent firms that received the permits free of charge. Moreover, emissions trading creates a new commodity market, which brings a range of interested parties to the table, including financial services companies, consultants, lawyers, potential sellers of credits, and so on. The effects of emissions trading on different business groups will be explored in more detail later in this chapter.

While environmentalists have generally favored command-and-control regulation, they support emissions trading, if given the choice between the two market-based instruments, mainly for three reasons. First, emissions trading guarantees the reduction of a certain quantity of emissions. Second, environmental groups are wary of pollution taxes because they make the cost of environmental protection explicit to the consumer. The potential cost arising from production under a tradable permit systems is passed on to consumers only indirectly. For this reason, industry might actually support emissions taxes in the hope that the public outcry about the costs leads to less stringent environmental standards (Stavins 1998). And third, internationalist environmental groups favor international emissions trading and in particular credit generation in developing countries because it offers a development mechanism. Critics consider these international distributional effects of emissions trading as north-to-south wealth transfers that should be disentangled from pollution reduction policies (Metcalf 2007).

Finally, legislators favor freely allocated allowances because the compliance costs are less visible than in the case of pollution taxes or auctioning permits. Allocating highly valuable pollution rights to industry also provides legislators with significant power over the distributional effects of environmental regulation. In the U.S. context, legislators might be especially interested in reducing the regional distributional effects of environmental regulation through control of the initial allocation of permits. The downside of the allocation of economic wealth in the form of permits through government officials is, William Nordhaus (2005) argues, that emissions trading is significantly more susceptible to corruption than pollution taxes.

In sum, emissions trading creates a number of incentives for business, environmental groups, and legislators to support the policy. Most notably, emissions trading creates a new commodity, which appeals in particular to business and government actors. The preference of key actors for emissions trading is strengthened by the fact that taxes are generally considered a difficult sell in the United States.[3]

Baseline-and-Credit versus Cap-and-Trade Systems

Two distinct forms of tradable permits systems exist: baseline-and-credit, and cap-and-trade (Ellerman, Joskow, and Harrison 2003). In baseline-and-credit systems, a regulatory body awards credits to those entities that overcomply with a regulatory standard (baseline). These credits can then be used to offset the surplus emissions of a noncompliant facility. Thus, baseline-and-credit systems serve to increase the flexibility in complying with a command-and-control environment (Hansjürgens 2005). In the Kyoto Protocol context, the clean development mechanism (CDM) and joint implementation (JI) are baseline-and-credit mechanisms, as will be explained later in this chapter.

By contrast, cap-and-trade systems set a limit (cap) for aggregate emissions, allocating a limited number of permits to regulated sources. Regulated entities must hold a permit for each unit of emissions, whereby they can trade permits among each other. In a cap-and-trade system, a government authority certifies the permits when they are initially issued. This is unlike baseline-and-credit systems, where credits are certified on a project-by-project basis.

Both types of emissions trading rely strongly on government regulation. Since scarcity is created artificially through government intervention, permit markets are regulation driven. The two forms of emissions trading evolved in a historical sequence. As will be further shown in the next section, the U.S. Environmental Protection Agency (EPA) first experimented with baseline-and-credit mechanisms before it implemented a cap-and-trade scheme.

The Intellectual and Political History of Emissions Trading

Emissions trading is a policy innovation that emerged in the United States as an alternative to command-and-control regulation and pollution taxes. U.S. economists developed the theoretical foundation of the instrument, and the EPA pioneered its application for a number of environmental pollution problems. Early experiments with emissions trading served to offer greater flexibility to companies in complying with a command-and-control regulatory environment. When the policy instrument matured, it offered a full-fledged market-based policy instrument to address environmental problems, with the Acid Rain Program being the reference case. Finally, in the mid-1990s, the instrument started to internationalize, in particular in the context of climate change politics. This global rise of emissions trading in climate politics is this book's central story, laid out in chapters 4 to 6.

U.S. Economists and the Idea of Emissions Trading, 1960s

"Economists spend much of their time attempting to understand how markets work. In the case of tradable permit schemes, the reverse is true: these markets have been created from theoretical considerations," pointedly comment Cédric Philibert and Julia Reinaud (2004) on the origins of emissions trading. Indeed, the history of the carbon market is first of all an intellectual one. The idea of emissions trading is widely seen as building on Ronald Coase's famous article "The Problem of Social Cost" (1960), which was a fierce critique of the Pigouvian tradition. Four decades earlier, in his book *The Economics of Welfare*, economist Arthur Pigou (1920) proposed to internalize economic externalities by means of taxation, which by 1960 was mainstream thinking in welfare economics. Coase challenged this notion by contending that participants should be left to negotiate the best possible solution for a situation where the social costs of economic activity arises.

Building on Coase's general theoretical argument, two scholars independently published the idea of a pollution market. Thomas Crocker (1966) considered a form of emissions trading for air pollutants in the United States, while John Dales (1968) explored emissions trading in the case of water pollution in Canada. Crocker and Dales maintained that auctioning off pollution rights would be the best way to arrive at the correct social value of pollution at the desired overall level of pollution. From the beginning, the debate was framed as emissions trading versus pollution taxes.

The early work was not clear about the purpose that emissions trading should serve. Economists discussed two notions (Gorman and Solomon 2002). The first idea suggested that emissions trading was a means to arrive at an optimal level and distribution of pollution to maximize social welfare. Consumers of clean air would have to compete with industrial players for the rights to use the resource. This would not necessarily lead to the optimal level of environmental protection, however. The second idea instead thought that emissions trading was meant to help achieve a predetermined level of emissions reduction in the most cost-effective way. The desired level of pollution reduction was the result of a political as opposed to a market process. The latter notion finally gained the upper hand.

W. David Montgomery (1972) delivered the first mathematical proof that trading permits could minimize the cost of achieving a predetermined level of pollution reduction. Montgomery considered three issues critical in designing a tradable permit scheme. Government needed to enforce limits on the overall pollution in a geographic space; a mechanism had to be devised for the allocation of permits to individual firms; and rules had to be defined for the trading of emissions rights between companies. This exploratory work by U.S. economists did not go much further into the details of designing emissions trading. The next step occurred when the EPA started to implement elements of emissions trading in U.S. environmental policy (Gorman and Solomon 2002).

U.S. Experience with "First-Generation" Trading Schemes, 1970s and 1980s

In the 1970s and 1980s, the first experiments with some form of emissions trading were being conducted under the Clean Air Act, in the phase-out of leaded gasoline and in the phase out of chlorofluorocarbons (CFCs). First-generation trading schemes were meant to provide for greater flexibility in a primarily command-and-control regulatory environment (Hansjürgens 2005). Mainly, first-generation trading systems were baseline-and-credit systems.

In the course of the Clean Air Act amendments in 1970, the National Ambient Air Quality Standards set air-quality standards for air pollutants including sulfur dioxide (SO_2), nitrogen oxides, and carbon monoxide. Emissions trading became an option for regions that could not meet the air-quality standards within the compliance period. In 1976, the EPA passed the Offset Interpretative Ruling for California, which said that new industrial plants in nonattainment areas (such as Los Angeles) must meet the lowest-achievable amount of emissions and offset their emissions elsewhere within the air shed. This method was included in the amendments of 1977 to the Clean Air Act. Thus, offsetting was introduced to U.S. environmental policy to achieve both environmental protection and industrial expansion. The requirement to obtain offsets for expanding production and pollution in a nonattainment area implied the possibility of trading offsets. Firms that reduced their emissions of air pollutants in a nonattainment area could theoretically sell offsets. The attempt at establishing an offset market in some areas in California failed due to low demand, poor verification and monitoring of emissions, and high transaction costs (Gorman and Solomon 2002). In the following years, the EPA introduced a number of additional emissions trading–related concepts under the Clean Air Act, which were met with suspicion by environmental groups. The new policy was seen to allow firms to escape emissions reductions.

Offsetting was also applied in phasing out and ultimately eliminating leaded gasoline in the United States. This was a case of indirect emissions trading, because in the first place participating entities traded production rights for leaded gasoline. In 1982, the EPA allowed refineries that had problems meeting their targets to purchase the right to use additional lead in production from refineries that exceeded their targets. While widely seen as a success, this program suffered from compliance problems, thus stressing the need for better monitoring systems.

The relative success of the phase out of leaded gasoline inspired the use of trading mechanisms in phasing out CFCs and other ozone-depleting chemicals. The 1987 Montreal Protocol on Substances That Deplete the Ozone Layer establishes binding timetables for the phase out of these substances, but does not specify policies or measures to implement the phase out. In 1988, the U.S. EPA issued tradable production allowances to producers and importers of ozone-depleting substances (Petsonk, Dudek, and Goffman 1998). Owing to the small number of producers, only modest trading activities occurred between the North American plants of DuPont and Dow Chemical between 1989 and 1996, when the production of CFCs was phased out (Lee 1996). Yet it is assumed that trading allowed the United States to achieve its Montreal targets at 30 percent less than the expected cost.

These three instances were first-generation experiments of emissions trading, all of which occurred in North America, mostly in the United States. At the time, the environmental policymaking paradigm was command-and-control regulation (Hansjürgens 2005). The breakthrough for market mechanisms came with the Acid Rain Program in the 1990s, which subsequently became the reference point for ideas of GHG emissions trading.

U.S. Experience with "Second-Generation" Trading Schemes, 1990s

Second-generation trading schemes are full-fledged cap-and-trade schemes that aim at replacing command-and-control regulation instead of increasing compliance flexibility within a command-and-control environment. The Acid Rain Program was the first nationwide cap-and-trade scheme. The program resulted from the amendments to the Clean Air Act in 1990. In the 1980s, scientific consensus emerged on the link between SO_2-emitting power plants fired by coal and oil, on the one hand, and the occurrence of acid rain, on the other. Efforts to reduce SO_2 emissions in the United States were soon caught in a stalemate between regions: the Midwest and Appalachian states versus the Northeast and Canada, dirty utilities versus clean ones. Most of the coal-powered electricity plants are based in the former regions, whereas the acid rain problem was prevalent in the latter due to the air transport of SO_2.

In an attempt to resolve the deadlock, senators Timothy Wirth (D-Colorado) and John Heinz (R-Pennsylvania) initiated Project 88: Harnessing Market Forces to Protect the Environment (Stavins 1988), a bipartisan effort to investigate the use of emissions trading as an environmental policy tool. They recruited Robert Stavins, a former economist with Environmental Defense and a professor at Harvard's John F. Kennedy School of Government, to direct the project. The final report recommended the use of emissions trading as a policy instrument for an array of pollution problems including acid rain and climate change. The rationale for Project 88 was to link environmental protection to the predominant economic logic in the 1980s. As Wirth put it, "Senator Heinz and I thought that economics was pervading everything else during the Reagan era and a lot of other issues were being looked at through an economic lens and why should environmental issues be excluded from that? . . . [E]nvironmental issues could not exist in a vacuum" (quoted in Bernstein 2001, 205)." While the idea of emissions trading had been discussed in economists' circles for decades, it was the bipartisan endorsement of the two senators that provided it with strong legitimacy in political circles. A *New York Times* article concluded: "Their imprimatur confers a new political legitimacy on economists' way of thinking about environmental problems" (Passell 1988). Project 88 is perceived to have been key in resolving the stalemate in the case of acid rain and more generally in promoting economic ideas in environmental policymaking (McCauley, Barron, and Morton 2008).

At the time, though, U.S. environmental groups such as the Natural Resources Defense Council were skeptical of emissions trading, seeing it as a "license to pollute." The success of the EPA's first-generation forays into emissions trading had been modest at best, and environmentalists felt that emissions trading was less an effective tool of environmental policy than it was a method to accommodate industrial growth (Buntin 1999). Only Environmental Defense stood apart from the environmental community in that it adopted a promarket, protrading strategy in the mid-1980s, as the instrument could be applied to acid rain and water conservation in Southern California (ibid.; Gorman and Solomon 2002). Fred Krupp (1986), the executive director of Environmental Defense, published an op-ed piece in the *Wall Street Journal* with the telling title "New Environmentalism Factors in Economic Needs" that advertised market mechanisms for solving environmental problems.

A number of key staffers in the White House under President George H. W. Bush were highly receptive to the ideas promoted by Environmental Defense and Project 88.[4] White House counsel C. Boyden Gray, Bob Hahn of the Council of Economic Advisers, assistant attorney general Richard B. Stewart for the environment and natural resources at the U.S. Department of Justice, and associate director Bob Grady of the Office of Management and Budget for Natural Resources, Energy, and Sciences were critical advocates of the approach (Gorman and Solomon 2002). In fact, Gray was the person who communicated Project 88's findings to the Bush team. He thought that a market-based approach could solve the gridlock over acid rain in Congress, because politicians would not decide who would win and who would lose; the market would.

On reading Krupp's article in the *Wall Street Journal*, Gray encouraged Environmental Defense to develop a market-based approach to the acid rain problem (Buntin 1999). In December 1988, he invited Krupp to the White House, asking Environmental Defense to draft a market-based acid rain policy for the Bush administration. While Environmental Defense initially hesitated to team up with a Republican administration because of the wider environmental community's reactions, it finally agreed to develop a policy draft. "The whole idea of emissions trading was to use this innovative, sort of Republican intellectual capital to do something more ambitious than would otherwise be possible," explains Krupp (quoted in Buntin 1999, 13). In February 1989, Environmental Defense unveiled its proposal, which subsequently fed into the interagency process of formulating acid rain policy. The administration ultimately decided to include emissions trading, proposing an overhaul of the Clean Air Act to Congress in July 1989. Congress approved a version of the bill in October 1990, submitting it two months later to the president for his signature. The first period of SO_2 trading in the United States ran from 1995 to 1999, with a second trading phase extending from 2000 to 2009. The program was widely seen as a success, because the environmental goals were met ahead of schedule and at 30 percent of the projected cost (U.S. National Science and Technology Council 2005).

The policy success of the Acid Rain Program triggered the spread of cap-and-trade schemes across the United States (Gorman and Solomon 2002). In 1994, California launched the Regional Clean Air Incentives Market (RECLAIM), which traded credits for reductions in nitrogen dioxide and SO_2 emissions. RECLAIM was not successful, in particular because it led to a concentration of pollutants in so-called hot spots—mostly urban and poor areas. The environmental justice movement, a "third force" in California environmental politics, thus became suspicious of the distributional effects of emissions trading. Nevertheless, other states followed suit in applying emissions trading. Illinois introduced a trading scheme to reduce smog-causing volatile organic compounds in the Chicago area. Eight states in the Northeast set up a regional trading market to control nitrogen oxides, which had been at the root of ozone-causing smog in the region.

Since the early 1990s, emissions trading has seen widespread adoption, especially in the United States, increasingly winning over a command-and-control approach to environmental policy. Next to the United States, only New Zealand had experimented with emissions trading prior to the big diffusion wave of emissions trading in climate politics. The Fisheries Act of 1996 established a quota management system for commercial fisheries in New Zealand (Petsonk, Dudek, and Goffman 1998). More recent applications of emissions trading include tradable renewable energy certificates, tradable energy-efficiency improvement certificates, waste management, and most important, tradable permits for GHG emissions (Philibert and Reinaud 2004).

International GHG emissions trading was first considered an option in climate policymaking in the run-up to the Rio Earth Summit of 1992, shortly after the United States had included trading in its Clean Air Act amendments. The first reports to promote the idea of international GHG emissions trading were *Project 88: Harnessing Market Forces to Protect the Environment* by Stavins (1988) and *The Greenhouse Effect: Negotiating Targets* by Michael Grubb (1989) from the Royal Institute of International Affairs. In this early phase, engagement with emissions trading in climate politics was a platonic affair for governments. Parties were not prepared to take binding targets and emissions trading on board. At the time, the first Bush administration was interested in exploring the applicability of market mechanisms to climate policy, as were international organizations such as the UN Conference on Trade and Development (UNCTAD). The White House organized a seminar on emissions trading in April 1990 (interview 42).[5] UNCTAD got interested in GHG emissions trading at an early stage in the process, as it saw the potential to leverage its experience in international trade in the climate politics field. Frank Joshua, the head of GHS emissions trading at UNCTAD, assumed a key role in driving UNCTAD's work in that regard. Prior to the UN Conference on Environment and Development (UNCED), the program convened a numbers of experts to propose an international emissions trading scheme. These experts included, among others, Grubb, who would later be a key adviser in the creation of the UK ETS and EU ETS, and Richard Sandor, the architect of the Chicago Climate Exchange (CCX), a voluntary U.S.-based market. UNCTAD (1992) published the report *Combating Global Warming: Study on a Global System of Tradeable Carbon Emission Entitlements*, for the first time outlining the notion of international GHG emissions trading.

The early attempts to promote emissions trading in international climate politics did not fall on fertile ground, however. The UNFCCC did not include emissions reduction targets, which made emissions trading obsolete. Nonetheless, the convention included one reference to what could be interpreted as the seed for what would later become emissions trading. According to articles 4.2(a) and 4.2(d), countries were allowed to achieve their emissions commitments jointly; that is, the excess emissions of country A could be offset by the emissions reductions of country B (Baron and Colombier 2005). Yet the article was ambiguous as to what joint implementation exactly meant. It wasn't clear if this referred to a cap-and-trade scheme, baseline-and-credit mechanisms, or some other form of international cooperation (Yamin 2005). The real rise of emissions trading in climate politics only began in the run-up to the Kyoto conference, which is the subject of the core of this book.

To sum up, until the internationalization of emissions trading through the Kyoto Protocol, the policy instrument had almost exclusively been implemented in the United States. The application of emissions trading in international climate policy signified a major step in the spread of emissions trading in several ways. First, the geographic scope was widened significantly, as emissions trading would be implemented—to a greater or lesser extent—across developed and developing countries with different national economies. Second, the organization of the market was no longer within nation-states at the regional or federal level, but at the international level. And finally, due to the ubiquity of GHGs in modern economies, the volume of tradable permits is many times higher than in previous trading experiments. Hence, in many ways, the rise of emissions trading in climate politics signifies a new order of magnitude of environmental markets. The following section draws the contours of the emerging global carbon market.

The State of the Global Carbon Market

The global carbon market is an elusive phenomenon. Major economies do not operate under emissions caps, and the future of carbon trading is up in the air, given that the first commitment period of the Kyoto Protocol ends in 2012. "Mind the [policy] gap" is the warning slogan of current market participants and policymakers. Despite the fragmentation of and uncertainty around the carbon market, though, the phenomenon is global, as we are witnessing the creation of trading schemes—both mandatory and voluntary—across a number of industrialized countries. In addition, some developing countries are involved in GHG emissions trading through credit-generating projects. The global carbon market includes both distinct forms of emissions trading: cap-and-trade schemes and baseline-and-credit trading though the CDM and JI. Yet the geographic scope of the trading activities and their financial scale vary significantly across the different forms of trading mechanisms.

Kyoto's Trading Mechanisms: Emissions Trading, the CDM, and JI

The Kyoto Protocol envisioned three distinct yet interrelated trading mechanisms: emissions trading, the Clean Development Mechanism (CDM), and Joint Implementation (JI). While the protocol created all three mechanisms, the operational details of the trading mechanisms were only agreed on in the Marrakesh Accord in November 2001.

The first trading instrument is known as emissions trading in Kyoto language, which inherently leads to terminological confusion, since the other two mechanisms are also forms of emissions trading. But in the Kyoto context, emissions trading only refers to a global cap-and-trade scheme as opposed to the two baseline-and-credit mechanisms, the CDM and JI. Article 17 of the Kyoto Protocol, which introduces emissions trading, allows industrialized countries—that is, those countries listed in annex I to purchase permits from other industrialized countries to increase its overall emissions cap. The Kyoto Protocol thus establishes a global cap-and-trade scheme for the period from 2008 to 2012. This form of emissions trading differs from national industry-based cap-and-trade schemes, such as the SO_2-trading program. First of all, the Kyoto Protocol binds governments as opposed to industrial entities, thereby establishing a government-based trading scheme. The protocol, though, allows treaty parties to authorize private actors to participate in any of the three trading mechanisms (Yamin 2005), based on the idea that it is more cost-effective to equate marginal abatement expenses across industrial sources than only across nations. Nevertheless, private sector participation in international trading is subject to the requirement that the country itself is eligible for international trading. Another striking difference between Kyoto's government-based trading scheme and domestic industry-based schemes is that the protocol does not establish financial sanctions in case of noncompliance (Baron and Colombier 2005). Given these differences, the transfer of the emissions trading concept from the domestic realm to the international system has raised critical questions of feasibility (see Victor 2001).

The CDM and JI are the two international baseline-and-credit trading mechanisms. Created by article 12 of the Kyoto Protocol, the CDM allows industrialized nations (annex I parties) to earn credits, or so-called certified emissions reductions, by implementing emissions-reduction projects in developing countries (nonannex I parties). Other approaches are that an entity in a developed country undertakes a CDM project unilaterally, or that an international financial institution such as the World Bank creates CDM projects. Every CDM project requires certification through the Executive Board of the CDM, an independent body under the UNFCCC, which also issues the certified emissions reductions. A critical and highly controversial issue in generating credits is "additionality." Article 12.5 provides that certified emissions reductions can be issued on the basis of "real, measurable and long-term benefits related to the mitigation of climate change; and reductions are additional to any that would occur in the absence of the certified project activity" (United Nations 1998, 12). Yet determining if a project is indeed additional is inherently difficult.

The political purpose of the CDM is twofold. First, the CDM allows firms in industrialized countries to exploit emissions reduction opportunities in developing countries that often are more cost-effective than those in industrialized countries that already have more efficient technology in place. Second, the CDM encourages nonannex I parties to contribute to CO_2 mitigation efforts, while not adopting binding targets and timetables. The CDM instead offers incentives to developing countries to engage in abatement efforts, as CDM projects result in a financial flow toward developing countries, and possibly in technological upgrades or leapfrogging.

As set out by article 6 of the protocol, JI allows industrialized countries (annex I parties) to generate credits, or so-called emissions reduction units, by jointly implementing abatement projects in industrialized countries. The JI mechanism is a hybrid between cap-and-trade and baseline-and-credit schemes, because it allows project-based emissions reductions and the transfer of permits within an international cap-and-trade scheme. Like CDM projects, JI projects require certification by independent certification bodies, which are overseen by the article 6 supervisory committee. The emissions credits generated through a JI project are deducted from the host country's amount of permits and added to the permit budget of the investing country. In sum, articles 17, 12, and 6 of the Kyoto Protocol laid out the trading mechanisms in theory, and the Marrakesh Accord specified the technical details. While theory is one thing, practice is another, as the evolution of trading activities in the aftermath of Kyoto demonstrated.

Mapping the Carbon Market: Scope and Scale

"In a single act, the Kyoto session created a highly ambitious agreement that requires a completely novel form of international financial trading to succeed, and no consensus on how to implement that scheme," David Victor (2001, 29) comments on Kyoto's top-down approach to establishing a global trading system for emissions permits. What followed in the decade after Kyoto was not a top-down implementation of a global trading scheme but rather a bottom-up process of trading experiments and schemes. Academics and market actors alike have been struggling to describe the nature of the emerging carbon market, with the key attributes being that it is fragmented (Tangen and Hasselknippe 2005), plurilateral (Sandor 2001), decentralized, and bottom-up (Victor and House 2004; Victor, House, and Joy 2005). From a theoretical international political economy perspective, the global carbon market has been conceptualized as a "regime complex" (Green 2008) and "political domain" (Betsill and Hoffmann 2008). A regime complex is "an array of partially overlapping and nonhierarchical institutions governing a particular issue-area"

(Raustiala and Victor 2004, 279). A policy domain, instead, is defined as "a component of the political system that is organized around substantive issues" (Burstein 1991, 328).

Victor, House, and Joy (2005, 1820) captured the underlying notion shared by these different attempts to describe and conceptualize the carbon market with the phrase "a Madisonian approach to climate policy": "The decentralized system is akin to the messy federalism that James Madison embraced in the U.S. Constitution. Whereas Madison foresaw individual states becoming *laboratories* for political innovation, this global federalism of climate policy has emerged through innovation within nations, regions, and individual firms." In the following, I draw a picture of the global carbon market by explaining its geographic scope (see figure 3.1) and distribution as well as by highlighting its financial scale. I strongly rely on two key reports that annually conduct this exercise of mapping the carbon market: the *Carbon* report of Point Carbon, a leader in market intelligence services related to the carbon market, and the *State and Trends of the Carbon Market* report of the World Bank.[6]

While the first trading experiments occurred as early as the late 1990s, the carbon market only took off in 2005, when the EU ETS was launched. In 2009, the global market reached a financial volume of $144 billion, with the EU ETS making up more than three-quarters of the pie (Kossoy and Ambrosi 2010). The carbon market thus remains the backbone of the market. Long-term projections of the carbon market's size are inherently uncertain, because volumes critically depend on domestic and international policy development, though it is likely that the financial market for GHG permits will be one of the largest. By setting emissions targets, the Kyoto Protocol allocated permits worth $2.3 trillion, based on assuming a price of $14 per ton of CO_2 (Victor 2001). Even if only a portion of these permits is traded globally, this results in a market of tremendous size.

The global carbon market can be segmented in different ways—for one, in terms of the distinction between mandatory markets, mostly resulting from Kyoto commitments, and voluntary markets. In 2007, the mandatory and voluntary markets held market shares by traded volumes of 98 and 2 percent, respectively (Point Carbon 2008). The mandatory, or compliance, market includes the EU ETS (61 percent of the total market by traded volume) as well as the CDM (35 percent) and JI (1 percent), the Kyoto Protocol's project-based mechanisms.

The EU ETS is the only existing multilateral trading scheme for CO_2 and the world's largest mandatory cap-and-trade scheme (chapter 5 focuses on the emergence of the scheme). As a *tributary* market to the Kyoto

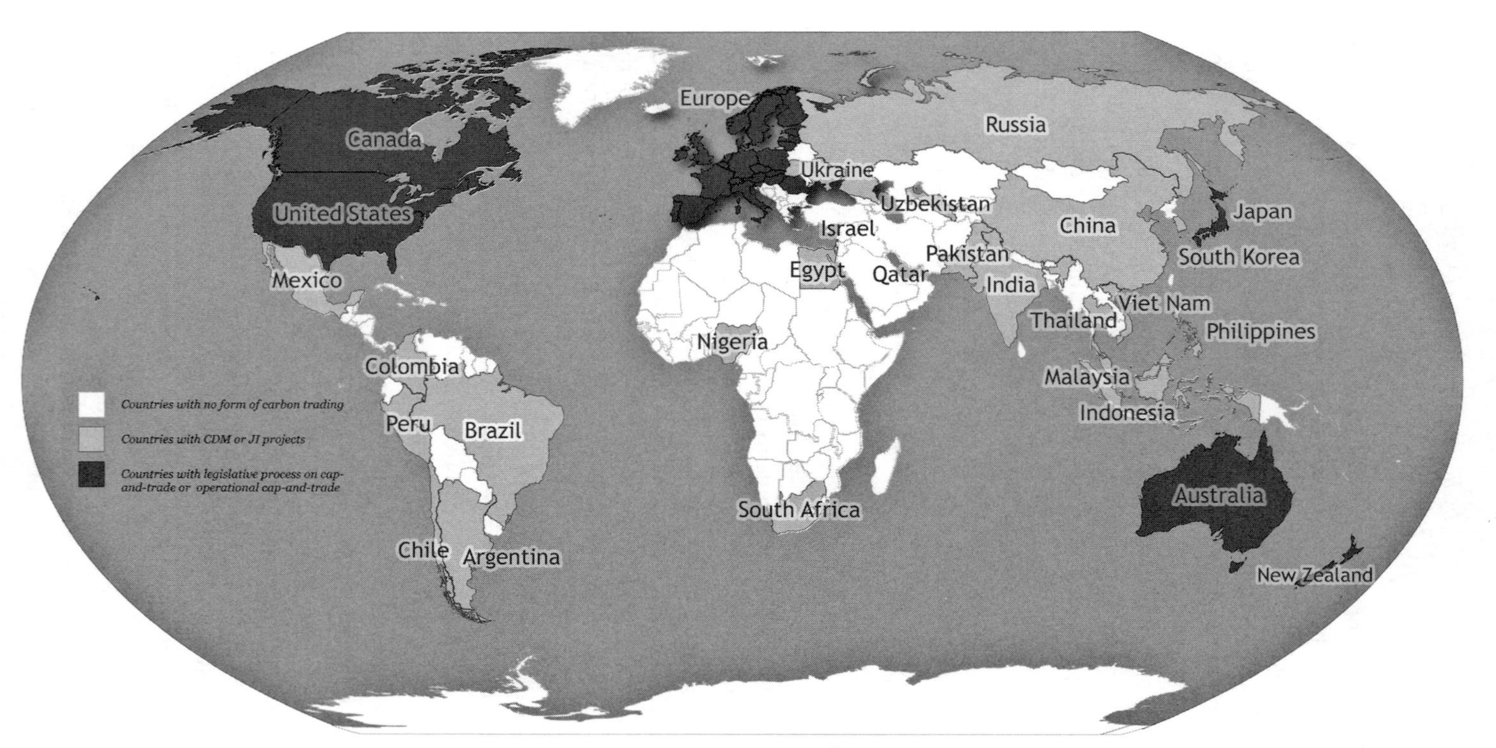

Source: Based on data from the CDM Executive Board (October 2010), http://cdm.unfccc.int/Statistics/Registration/AmountOfReductRegisteredProjPieChart.html. The map only shows countries with CDM projects that generate more than 1 million expected average annual CERs.

Figure 3.1
The global scope of carbon trading

Protocol, the scheme serves to achieve the EU's Kyoto target. Legislation for the EU ETS was adopted in 2003, and actual trading began in January 2005, with a pilot phase running until 2007. The second trading period of the EU ETS runs parallel to the first commitment period under the Kyoto Protocol from 2008 to 2012. In spring 2007, the EU heads of state decided to ensure the long-term continuity of the EU ETS by setting an emissions reduction target for 2020. Also in 2007, the EU ETS underwent a review process, which led to an institutional overhaul of the scheme, including issues such as sectoral coverage and allocation method. The second-largest GHG market after the EU ETS is the New South Wales Greenhouse Gas Abatement Scheme in Australia. Covering only the electricity sector, the New South Wales scheme was launched in 2005, reaching a value of about US$117 million in 2009 (Kossoy and Ambrosi 2010).

Since the EU ETS is the major mandatory trading scheme, it is also the main driver for project-based mechanisms by creating a demand for credits.[7] In 2008, the primary CDM market was worth $6.5 billion (ibid.). Early projects produced credits by reducing industrial gases that had an especially high global warming factor such as HFC-23 and N2O, considered to be low-hanging fruit. Since 2007, more credits increasingly result from renewable energy and energy-efficiency projects. Since the inception of the CDM market, China has been the largest seller country, generating 72 percent of all CDM credits in 2009 (Kossoy and Ambrosi 2010). The second-largest seller was Africa with 7 percent, followed by Asia, Brazil, India, and others, each with less than 5 percent of the market share. On the buyer side, the United Kingdom held the largest share with 37 percent of the market, followed by Japan with 13 percent. Given that London is the financial center of the carbon market, the United Kingdom has been an active trader. Furthermore, the World Bank has assumed an active role in the CDM market, pioneering the development of a number of carbon funds, starting with the launch of the Prototype Carbon Fund in 1999 (Cooper 1999b).

Compared to the CDM market, the JI market is considerably smaller, with most credits being generated in Russia and Ukraine. Emissions reductions were achieved mainly through gas leakage abatement, nitric acid, landfill, and fuel-switching projects. The EU ETS, in short, is the backbone of the current mandatory market. Through credit-based mechanisms, the EU ETS drives the development of credit-generating activities in China, Russia, Ukraine, and a number of developing countries.

Representing the second pillar of the carbon market, the voluntary market is a credit-based trading market in which credits are generated and sold for noncompliance purposes. In 2009, the voluntary market had only a financial volume of $338 million, which is miniscule compared to the total size of carbon markets (Kossoy and Ambrosi 2010). With a volume of $143 million, the United States accounted for about 40 percent of the voluntary market in 2009. Buyers in the voluntary market are companies and consumers voluntarily offsetting their GHG emissions. Voluntary carbon offsets are often marketed under the "carbon neutrality" label. Moreover, voluntary trading mechanisms, such as the CCX serve as models and dry runs for emerging compliance markets. A key issue for the voluntary market is the quality of emissions reduction credits, which led to the creation of a number of standards, including the Voluntary Carbon Standard and the Gold Standard (Newell and Paterson 2010). Both standards certify emissions credits in the voluntary markets. While the Voluntary Carbon Standard was introduced by industry, the Gold Standard is primarily an initiative of environmental groups.

The global carbon market, consisting of compliance and noncompliance markets, has been growing considerably since 2005. Yet in terms of geographic and sectoral coverage as well as volumes traded, the existing market remains highly fragmented and has not yet led to significant CO_2 abatement. The political significance of the rise of emissions trading goes beyond the real carbon market, and includes the many political and legislative attempts to establish cap-and-trade schemes, most notably in the United States.

Trading Schemes in the Making: North America and Asia Pacific

The political domain of emissions trading includes multiple activities that serve to prepare trading schemes, before they have led to actual trading. A number of subfederal entities and nation-states are moving toward adopting one or the other form of emissions trading. These activities are concentrated in North America, Australia, and Japan, basically covering the triad plus Australia and Canada. All of these countries committed to national emissions reduction targets in the Copenhagen Accord of December 2009.

The highest level of activity has been observed in the United States at both the state and federal levels. Chapter 6 explicitly deals with the United States, which is why I offer only a primer here. Ten states in the northeastern United States created the Regional Greenhouse Gas Initiative (RGGI), which established a regional cap-and-trade scheme for the power sector, due to begin operations in 2009. In 2006, California adopted the Global Warming Solutions Act, which set an aggressive emissions reduction target for the entire state's economy. The administration aims to achieve the target through emissions trading, which has to be implemented by 2012. In February 2007, five Western states set up the Western Climate Initiative, with the goal of developing a regional multisector trading scheme.[8] A few months later, Utah and the Canadian provinces of British Columbia and Manitoba joined the initiative. The Midwestern states along with the Canadian province of Manitoba signed the Midwestern Greenhouse Gas Accord in November of the same year.[9]

With U.S. states taking the lead, federal U.S. politics has woken up to mandatory climate policies. Since 2003, a number of market-based climate bills have been proposed in Congress. In 2009, the House of Representatives passed the first comprehensive cap-and-trade bill, the American Clean Energy and Security Act. The bill was proposed by Congresspeople Henry Waxman and Ed Markey, and was a major milestone. Yet at the time of writing this book, the legislative process on cap and trade has stalled. The Senate abandoned the aim of passing legislation in 2010.

While the United States is not party to the Kyoto Protocol, its neighbor is. In April 2007, Canada's government announced a new climate strategy, setting an absolute emissions reduction target of 20 percent by 2020 based on 2005 levels. For industry, the plan sets the intensity target of reducing emissions 18 percent by 2010 for every unit of production (Government of Canada 2008). These targets can be implemented through a number of mechanisms including interfirm trading. While the scheme will initially be domestic, Canada is exploring linkage with the United States and Mexico. The plan provides for offsets from both nonregulated sectors of the Canadian economy and CDM projects.

A similar trend toward adopting domestic emissions trading schemes can be observed in the Asia Pacific region, particularly in New Zealand, Australia, and Japan. In September 2008, New Zealand passed legislation on the New Zealand Emission Trading Scheme. The future design of the scheme was laid out in the Climate Change Response Amendment, passed in November 2009. The implementation of carbon trading started with the forestry sector in January 2008 and subsequently included other sectors in the program. The scheme is supposed to cover all sectors by 2015. New Zealand is now the first country outside Europe that has a mandatory, economy-wide emissions trading scheme. Compared to New Zealand, the legislative process on cap and trade in Australia has been less successful to date. In May 2007, the prime minister's Taskforce on Emissions Trading submitted its final report, recommending a national emissions trading scheme. In July 2008, the government published a "green paper" laying out the scheme's future design. The implementation of the plan for a Carbon Pollution Reduction Scheme, an economy-wide emissions trading scheme, was postponed in April 2010, however, although it will be reexamined at the end of 2012 (Kossoy and Ambrosi 2010). This measure was due to domestic political deadlock and the lack of international progress on climate policy.

Though host to the birthplace of the Kyoto Protocol, Japan has always been skeptical of emissions trading as a policy instrument. In October 2008, Japan's government launched a voluntary trial carbon-trading scheme, which is supposed to pilot a mandatory cap-and-trade scheme (Maeda 2008). The scheme covers about 70 percent of the CO_2 emissions from industry. In March 2010, the Japanese government proposed the Basic Act on Global Warming Countermeasures, which foresees a mandatory cap-and-trade scheme, carbon tax, and feed-in tariff for renewable energy sources (Kossoy and Ambrosi 2010). It represents a bold shift toward market-based climate policy in Japan. As of summer 2010, it is still unclear how the politics of passing the legislation will unfold. The Tokyo metropolitan area, meanwhile, launched its own cap-and-trade scheme for office and commercial buildings in April 2010. Other countries considering national emissions trading schemes include Mexico and South Korea (Point Carbon 2010)

In addition to the various national initiatives, there have been activities to integrate the fragmented carbon market. In October 2007, the European Commission, a number of EU member states and U.S. states, Norway, and New Zealand launched the International Carbon Action Partnership.[10] This partnership is designed as a policy forum to advance the linkage between regional trading schemes and assist countries in designing emissions trading schemes. Altogether, the portrayed initiatives indicate a trend toward the proliferation of national and regional emissions trading schemes. Furthermore, efforts are being undertaken to integrate the trading schemes that have sprung up in a "Madisonian" way. A Point Carbon survey showed that two-thirds of the respondents expect a global reference price for CO_2 emissions by the year 2020 (Point Carbon 2010). Having outlined emissions trading as a theoretical, historical, and current phenomenon, I now turn to the distributional effects of climate policy and emissions trading on business.

Business and Carbon Trading: Risks and Opportunities

Industry is not only part of the problem but also part of the solution to climate change. Accordingly, some firms face predominantly business risks from climate policy, while others stand to benefit from it (Pinkse and Kolk 2009). Broadly speaking, the higher the GHG intensity of a firm, the higher are its climate risks, meaning the higher are its compliance costs. The group of emitters, comprising the oil and gas industry, the electricity sector, and energy-intensive manufacturing industries, bears higher costs than benefits from climate legislation overall. Climate policy instead offers business opportunities to low-carbon technology providers, such as renewable energy companies and nuclear energy producers, as well as permit sellers, traders, and other intermediaries in the carbon market. Hence, for the second group of firms, the benefits of climate regulation prevail over the costs. Below I discuss the corporate winners and losers in climate politics, and their respective political strategies, thereby helping to illustrate the lines along which competing carbon coalitions emerged. In doing so, I draw on the typology of corporate political strategies in environmental politics developed in chapter 2.

Managing Risks: Oil, Power, and Manufacturing Industries

The consumption of fossil fuel–based energy is pervasive in modern economies, resulting in a high number of sources of GHG emissions. The largest industrial emitters are found in the oil and gas industry, electricity sector, and energy-intensive manufacturing industries. To them, climate policy, and in particular carbon pricing, is a risk: "To the extent that society starts pricing these various externalities, that is a negative for the valuation of the polluting firm" (Llewellyn 2007, 9), a report by the investment firm Lehman Brothers stated. While in the short term climate policy raises the cost of fossil fuels as compared to other energy sources, in the long term it might lead to a shift in energy sources. Thus, a static economic analysis suggests that fossil fuel–based industries resist climate policy (antiregulatory strategy). The early and unified opposition of industry through the Global Climate Coalition (GCC) was indeed largely driven as well as organized by the oil industry, coal-based power companies, and energy-intensive manufacturing industries, such as the chemical sector and automobile industry (Levy and Egan 1998; Levy and Rothenburg 1999; Goel 2004).

While the big picture suggests that regulatory costs prevail over benefits for emitters, the costs vary strongly across emitters. For instance, a study on the CO_2 emissions in relation to the revenue of a number of European airlines showed a high variation across the sample: one firm's emissions were 64 percent above the average, while another firm emitted 20 percent less than the average (Llewellyn 2007). Furthermore, a study of thirty-three U.S. electricity companies revealed that the rate of return adjusted for the damage value of emissions (CO_2, SO_2, and nitrogen dioxides) ranged from -14.2 to 6.7 percent (Repetto and Dias 2006). By that calculation, the environmental costs caused by American Electric Power, a large coal-based electricity generator, exceeded its surplus, resulting in a net added value of minus $4.8 billion. These examples demonstrate that there are considerable intraindustry differences in carbon intensity, and that these result in different compliance costs for firms.

Though the baseline policy preference is no emissions controls and the baseline strategy is opposition to carbon controls, companies from fossil fuel–based industries have also advocated voluntary emissions reductions and emissions trading with the free allocation of permits along with generous targets (proregulatory risk management strategy). This is mostly due to shifts in the institutional environment. In the second half of the 1990s, the policy preferences of U.S. and European oil firms started to diverge with European oil majors advocating emissions trading as the most cost-effective mandatory climate policy. U.S. majors instead continued to reject binding emissions controls. Yet the production portfolios of European and U.S. oil and gas companies do not vary significantly since their core business firmly remains oil and natural gas recovery and sales (Skjaerseth and Skodvin 2001, 2003). Although BP and Shell have aggressively advertised their investments in renewable energies, this does not significantly affect the carbon intensity of their businesses. In fact, BP is more carbon intensive than ExxonMobil, the often-blamed culprit in climate politics, if the ratio of natural gas to oil reserves is taken as an indicator (Rowlands 2000). Natural gas is generally considered a *transition fuel* that is less carbon intensive than oil.

The only aspect regarding market positioning that could explain different attitudes toward carbon trading is the extent to which the firms have well-established trading businesses. Shell is involved in oil and gas trade to a much greater extent than ExxonMobil. With economic effects playing a marginal role for the different policy preferences of European and U.S. oil firms, scholars have frequently pointed to home country institutions as a key explanatory factor. Different levels of regulatory pressure and public opinion have been the main drivers for the split of the oil industry over political climate strategies (Levy and Newell 2000; Levy and Kolk 2002; Skjaerseth and Skodvin 2003; Pulver 2004). The strong public demand for climate action in Europe led firms to advocate low-cost forms of climate policy such as emissions trading.

In the power industry, differences in political strategies strongly correlate with the fuel mix of electricity companies. The higher the GHG emissions of a utility, the higher its climate risk (Gardiner 2006). Hence, under a carbon-pricing scheme, coal-based electricity generators are more risk exposed than utilities with strong assets in nuclear energy and natural gas, providing them with a competitive disadvantage. Nuclear power generators are instead more likely to accept a price on carbon than coal-based utilities, because climate policy might offer nuclear power generators a competitive advantage. When the institutional pressure for emissions controls has increased, though, even some coal-based utilities in Europe and the United States have come out in support of emissions trading. Hence, if regulation is inevitable, utilities might lobby for carbon trading with grandfather clauses—that is, the free allocation of permits and a weak cap to contain compliance costs.

The group of energy-intensive manufacturing industries includes a number of industries, such as the chemical, automobile, paper, steel, cement, and aluminum sectors. As energy costs are a high share of the overall production costs in these sectors, they are vulnerable to carbon-pricing policies. If a manufacturing company has taken the industry lead to increase the energy efficiency of its production process and reduce emissions, however, it has a case to support emissions trading. Given that credits for early action are allocated freely, the company can sell permits due to its early emissions reduction efforts. This gives rise to conflict between technological leaders and laggards.

Many energy-intensive manufacturing companies are also multinationals competing in a global marketplace. Therefore, one of their biggest concerns regarding climate policy is how unilateral emissions controls affect their international competitiveness. This can lead to different political strategies. Companies can oppose mandatory climate policy in general (antiregulatory strategy). Alternatively, they can push for weak emissions reduction targets or a global climate regime to create a level playing field with competitors, thus "trading up" climate policy (Vogel 1995; DeSombre 2000) (proregulatory risk management strategies).

Seizing Opportunities: The Carbon Industry, Technology Firms, and Investors

On the other side of the equation are a number of industries for which climate regulation in general and carbon trading in particular offer business opportunities. Hence, for these firms the benefits of regulation outweigh the compliance costs. This is because carbon trading is a market-creating policy resulting in the creation and growth of primary and secondary markets. The market for emissions permits is a direct product of emissions-trading policies, which is why I refer to it as a primary market. The carbon market creates opportunities for permit sellers (regulated entities with excess permits or project developers) as well as market intermediaries such as financial services, accounting businesses, and law firms. Secondary markets are those markets that experience growth due to the introduction of a price on carbon, including markets for renewable energy and other low-carbon technologies, but also the market for natural gas.

The market-making effects of climate policy have been gaining ground in political debates since the late 1990s. Embedded in discourses on sustainable development and ecological modernization, ideas on green growth have become popular. A number of countries such as Germany and the United Kingdom have bought into these in an effort to develop lead markets for environmental technologies and renewable energy (Jacob et al. 2005). In fact, Karine Matthews and Matthew Paterson (2005) consider market creation a main driver behind climate policy. Emissions trading, project-based credit generation, and renewable energies can all be interpreted as new sites of accumulation. Lured by the prospect of market creation or expansion, a great number of firms have joined the political game. The emergence of market-based and market-creating policies goes hand in hand with firms seeking opportunities in environmental markets: "This shift in government policies is also reflected in a shift of position by the regulated firms. While firms' behaviour under command-and-control regimes consisted of cost-avoidance strategies, the shift towards incentive-based instruments is mirrored by companies exploiting the environment in search of markets for new products to gain a competitive edge" (Carraro and Egenhofer 2003, 5).

Next to the regulated entities that act as sellers and buyers of permits, a number of other businesses and services stand to benefit from carbon-trading policies: project developers generate credits through CDM projects; exchanges organize market transactions; investment banks set up carbon funds and financial risk management products; lawyers create the legal framework of the new global market; accounting firms verify emissions; and consultancy firms offer specialized services such as risk management (Labatt and White 2007; Pinkse and Kolk 2007, 2009). These actors are attracted by the business opportunities of a new and large financial market. They strongly favor emissions trading over any other form of climate policy because their core business depends on the existence of emissions trading schemes. Additionally, they have an interest in long-term legal frameworks and tight caps, which create scarcity—a prerequisite for a well-functioning market. To date, the carbon industry is based largely in London, which successfully pursued its strategy to become the hub of the carbon market.

Compared to the level of the emitters' political involvement, the carbon industry only engaged with climate policy at a relatively late stage when it was clear that emissions trading was the future. As will be shown, the beneficiaries of carbon trading initially pursued a nonparticipation strategy. Only when a strong momentum for carbon trading existed did they begin to engage in advocacy, adopting a proregulatory market-making strategy. While both the emitters and the carbon industry are represented by the International Emissions Trading Association (IETA), the financial intermediaries also created their own trade associations, including the Carbon Market Association in the United Kingdom, the London-based International Carbon Investors and Services, and the Washington-based Environmental Markets Association. The political goal of these associations is the establishment of well-functioning trading schemes along with their global integration. These associations have been crucial sources of expert advice in the creation of a number of trading schemes.

Next to the carbon industry, providers of environmental technologies benefit from climate policy. Renewable energy companies, providers of energy-efficient technologies, and also the nuclear industry gain competitive advantages through carbon-pricing policies. So-called clean-energy markets are on the upswing: global investment in all clean-energy sectors reached $148 billion in 2007, which is 60 percent above the $93 billion tracked in 2006 (Makower, Pernick, and Wilder 2008, 6). As regards policy preferences, technology providers are often indifferent as to whether a carbon tax or emissions trading is chosen. What matters is the price of carbon that makes their products more competitive. Yet frequently carbon-pricing policies do not initially lead to prices high enough to stimulate investment in low-carbon technologies. Therefore, many technology companies have a keen interest in subsidies that support the takeoff of their technologies. Key business groups representing low-carbon technology providers include a number of technology-specific associations (solar, wind, nuclear, and so on), which over time consolidated into larger associations. The renewable energy business has always had a voice in climate politics through associations such as the U.S. and European Business Councils for Sustainable Energy (Dunn 2005). As they represented a nascent industry, though, their political clout was limited. Only when multinational technology companies such as General Electric (GE) awoke to the climate issue did the sector gain political influence. Similar to the carbon industry, large technology providers first refrained from advocacy activities before they started to pursue a proregulatory market-making strategy.

Apart from the carbon industry and technology companies, reinsurers and institutional investors are in support of stringent climate policy. For reinsurers, the primary reason for adopting a progressive political strategy is not to create or expand markets but rather to protect their business from the financial risks of natural disasters related to climate change, such as severe weather events and floods. The reinsurers got involved in climate politics from the mid-1990s through the Insurance Industry Initiative of the United Nations Environment Programme (UNEP), which in 2003 merged with the UNEP Financial Institutions Initiative into the UNEP Finance Initiative. Though the reinsurance industry extensively engaged with the topic, it never threw its weight behind policy proposals to mandate emissions cuts, disappointing environmentalists who had hoped to have found an industry ally in the push for climate policy (Leggett 2001). The reasons for the reinsurance industry's reluctance to become a vocal advocate are related to the broader political economy of the insurance industry. Given a market of $1.4 to $2 trillion, the insurance industry seems to hold tremendous economic power. Yet Paterson (2001) argues that the insurance industry does not occupy critical nodes in the financial system, which makes it economically and politically less powerful than its market size suggests.

Beyond the group of insurers, institutional investors have come to play a role in climate politics by calling for action. They are neither clearly on the winner nor the loser side of the distributional game of climate politics, since their investment portfolios include assets in a large variety of economic sectors and companies. Hence, they aim to strike a balance between managing the risks and seizing the opportunities of climate policy. A number of investor groups have formed that engage with climate policy and promote the climate risk disclosure of companies, including the Investor Network on Climate Risk, the Carbon Disclosure Project, and the Institutional Investors Group on Climate Change.

Conclusion

This chapter outlined the political economy of emissions trading by considering different types of trading, its intellectual and political history, the current carbon market, and its effects on industry. Since the late 1980s, market-based environmental policy has been on the rise, outpacing command-and-control regulation. In the battle for the supreme market-based environmental policy, emissions trading has frequently won over pollution taxes. Developed by U.S. economists in the 1960s, emissions trading was first implemented by the EPA at the regional and later the national level in the United States. Early experiments with emissions trading introduced baseline-and-credit mechanisms, while the Acid Rain Program created the first full-fledged cap-and-trade scheme. Emissions trading thus has been incubated in the U.S. economy.

With the application of emissions trading to climate policy, the internationalization of emissions trading began. Though the policy instrument had been introduced top-down through interstate diplomacy in the Kyoto Protocol, the actual global carbon market emerged in a bottom-up approach, including national and regional cap-and-trade programs, voluntary trading activities, and the CDM and JI projects. The result is a fragmented and plurilateral trading scheme that is slowly moving toward global integration. To date, trading activities revolve around the EU ETS, although most other industrialized countries in North America and Asia Pacific are moving toward adopting cap-and-trade schemes, intending some form of linkage to the international permit market.

The rise of emissions trading in climate politics has considerable distributional effects on industry, which critically influences how different business groups engage with climate policy and along which lines coalitions form. Inter- and intrasectoral divides exist regarding policy preferences and strategies. Carbon-intensive industries—including the oil, gas, coal, and energy-intensive manufacturing industries as well as power generators—face significant risks and costs from emissions reduction mandates. This had led them to adopt mainly two types of strategies depending on the level of institutional pressure: with a low level of public demand for climate action emitters tended to oppose climate regulation (antiregulatory strategy); and under high institutional pressure emitters instead adopted voluntary commitments and/or advocated emissions trading (proregulatory risk management strategy). The prospect of the free allocation of emissions reduction permits was especially an incentive that led big emitters to accept and support emissions trading.

On the other hand, carbon-pricing policies and carbon trading in particular offer opportunities to financial services, technology providers, and investors, as climate policy creates a primary market, the carbon markets, and expands secondary markets, such as for low-carbon technologies. The winners of climate policy primarily responded in two ways: under low institutional pressure firms tended to refrain from political engagement (nonparticipation); and under high institutional pressure firms promoted market-making policies (proregulatory market-making strategy).

As will be demonstrated, the protrading coalition essentially pursued a proregulatory risk management strategy, as it mainly gathered big emitters. The following case studies turn to three decisive historic steps of the global diffusion of emissions trading: the internationalization of a U.S. regulatory approach in the Kyoto Protocol; the U-turn of the EU from skeptic to champion of carbon trading; and the return of carbon trading to the United States. These three steps reflect the path of diffusion of carbon trading. After emissions trading had been included in an international agreement, the EU was the first actor to develop a trading scheme. Thereafter, U.S. states started developing cap-and-trade schemes. I will analyze what role the protrading business lobby played in the rise of carbon trading and what its sources of influence were. The book now turns to the first part of the story: how emissions trading, a U.S. regulatory instrument, came to be included as a major pillar of the Kyoto Protocol.

4

The Kyoto Protocol: Internationalizing a U.S. Regulatory Approach

At the second Conference of the Parties (COP 2) in Geneva in 1996, Timothy Wirth, the undersecretary for global affairs in the U.S. State Department and head of the U.S. delegation, said: "The US recommends that future negotiations focus on an agreement that sets a realistic, verifiable and binding medium-term emissions target . . . met through maximum flexibility in the selection of implementation measures, including the use of reliable activities implemented jointly, and trading mechanisms around the world (quoted in Grubb, Vrolijk, and Brack 1999, 54)." With these words, the United States put international carbon trading on the negotiation table of global climate politics for the first time. Not long after that, in December 1997, the Japanese city of Kyoto became the birthplace of the international treaty, which foresaw international emissions trading as the key implementation mechanism for international climate policy. Prior to 1997, emissions trading was a U.S. regulatory approach that had been successfully implemented on a small scale with a limited number of actors in the Acid Rain Program and other domestic programs (Gorman and Solomon 2002). The Kyoto Protocol internationalized the regulatory instrument, initiating a process of global diffusion, and a number of private experiments with emissions trading and so-called son-of-Kyoto bills emerged around the globe.

This sudden rise to prominence of market mechanisms has been an enigma to observers. It is clear that U.S. foreign policy in the run-up to Kyoto was decisive for the inclusion of flexible mechanisms in the protocol. Yet it remains inconclusive as to why the United States suddenly embraced binding targets under the condition that international flexibility applied, while it had rejected binding targets before. I argue that a change in the transnational associational politics of business occurred in the mid-1990s, which—together with a new administration—led to a shift in foreign policy. While in the early 1990s business stood united in its opposition to climate policy by fighting NGOs, an NGO-business alliance emerged in the mid-1990s that supported emissions trading as the new climate compromise.

This chapter traces the shifts in the associational politics of business, and their effects on climate policy and the rise of market mechanisms in the 1990s. First, I show how a strong antiregulatory business coalition successfully fought international carbon constraints in the early phase of climate politics. Second, I contend that in the mid-1990s, business conflict emerged and an NGO-business coalition formed. This alliance critically influenced U.S. foreign policy and the outcome of the Kyoto conference, which led to the internationalization of emissions trading. Third, I describe how this proemissions-trading lobby institutionalized in the post-Kyoto period, while the antiregulatory network started to disintegrate.

The Business Opposition in Control of Early Climate Politics, 1989–1995

In 1988, the Intergovernmental Panel on Climate Change (IPCC) was created under the auspices of UNEP and the World Meteorological Organization. The IPCC became the scientific body that provided knowledge to the diplomatic process of international climate politics. Two years into its existence, the IPCC produced its first report, which said that the rise of global temperatures was caused by rising concentrations of GHGs in the atmosphere due to human activities (cited in Grubb, Vrolijk, and Brack 1999). In February 1991, in the run-up to UNCED, the international community started negotiating what came to be the UNFCCC. Climate change had matured from a scientific matter to a political issue. The heat was on.

The Rise of Organized Business Opposition

Since the inception of climate politics, industry played a pivotal role, critically influencing international and domestic policy. Up until the agreement of the Kyoto Protocol, the major forums for collective action of industry were the GCC and a few allying organizations, including the American Petroleum Institute (API), the International Chamber of Commerce (ICC), and the Climate Council (Goel 2004).[1] These groups organized transnational and transatlantic industry opposition against climate regulations, with U.S. firms being instrumental in the process. Until the mid-1990s, this antiregulatory business coalition was highly successful in averting international and domestic emissions reduction obligations.

The GCC was founded in 1989, and represented about 40 percent of the U.S. economy and several multinational companies (Pulver 2005). Members came mainly from the fossil fuel and energy-intensive manufacturing industries, including the oil, coal, automobile, electricity, cement, aluminum, steel, chemicals, and paper industries (Levy and Egan 2003). The GCC was mainly organized by the U.S. oil industry, which has—compared to other business interests—extraordinary influence in shaping U.S. domestic and foreign policy on energy-related issues (Goel 2004).[2] While both Exxon and Texaco were official board members, BP and Shell were regular members through their foreign subsidiaries. Exxon was especially influential in the API, which again was a core member of the GCC (Pulver 2004). The GCC pursued two main strategies to influence climate policy in the 1990s (Levy and Kolk 2002; Goel 2004). For one, it attempted to discredit climate science through an information campaign. Also, once scientific consensus became less questionable, the GCC framed its opposition to climate regulations as a matter of economic costs and a question of the participation of developing countries.

The influence of U.S. firms in organizing the global industry position extended to other global and regional business organizations as well. Until 1998, a representative from Texaco was the ICC's chair. Exxon's Brian Flannery, a climate scientist and key lobbyist, played a pivotal role in the climate change work of the International Petroleum Industry Environmental Conservation Association, the London-based association of the oil industry dealing with environmental issues (Leggett 2001). Finally, Exxon strongly influenced the environmental work in the European Petroleum Industry Association (EUROPIA), the industry organization of oil firms operating in Europe. Altogether, a small number of business associations coordinated the corporate political activities of global industry. U.S.-based firms in general and Exxon in particular were dominating the strategic course of the collective action of critical industries (Pulver 2005). It is reported that Exxon alone funded the GCC in the 1990s with seven hundred million pounds (Skjaerseth and Skodvin 2003, 49).

Preventing International and Domestic GHG Regulation

The first major battle between industry and environmentalists was fought over the UNFCCC, which was agreed on at the Earth Summit in Rio de Janeiro in June 1992. The convention set the nonbinding goal of stabilizing GHG emissions at 1990 levels by 2000. It did not contain any binding emissions reduction targets and timetables. This was particularly due to

U.S. opposition against mandatory targets under the Bush Sr. administration. The U.S. negotiating position was largely influenced by U.S. fossil fuel interests, which enjoyed good access to policymakers under the Republican administration (Skjaerseth and Skodvin 2003). Moreover, White House chief of staff John Sununu was skeptical of the science of climate change, which lent support to antiregulatory business interests from within the administration (interview 42). In the end, the UNFCCC did not pose a regulatory threat to the global oil industry because it did not contain clear commitments on the side of governments.[3] The API was largely pleased with the agreement.

In the UNFCCC's aftermath, the venue shifted to the domestic level, as both the United States and the EU were promoting national climate policies. In the United States, oil and gas firms were facing a similar situation as their European industry peers. After President Bill Clinton took office in January 1993, he announced the U.S. target of reducing GHG emissions to 1990 levels by 2000. In support of this stabilization target the administration proposed a British thermal unit tax, which was supposed to be based on the heat content of the fuel. The tax was rejected by the Senate, which at that time had a Democratic majority. The U.S. oil industry played an important role in killing the tax proposal. The API and other associations formed the American Energy Alliance, which was advised by the public relations firm Burson-Marsteller and targeted individual Democrat senators. The API alone paid US$1.8 million for a grassroots letter and phone-in campaign against the tax (Newell 2000, 100). This case was indicative of the strong influence that the oil industry exerted on Congress throughout the history of U.S. climate politics (cf. Goel 2004). The Republican Party's landslide victory in 1994 further strengthened the influence of industry groups in Congress. As a consequence of its defeat in the case of the carbon tax, the Clinton administration became increasingly inaccessible to the oil industry and instead consulted more closely with the environmental movement (Skjaerseth and Skodvin 2003). The political cleavage increasingly ran between business, on the one hand, and environmental groups and the administration, on the other.

A similar battle was fought in the EU. Preparing for the Rio conference, the European Commission proposed a package on climate policy including a carbon-energy tax in 1991. European industry associations spearheaded by Business Europe (formerly the Union of Industrial and Employers' Confederations of Europe, or UNICE), the umbrella organization of thirty-four business associations, and EUROPIA (ibid.) fiercely opposed this tax. The latter rejected any new tax on fossil fuel products. The lobbying campaign proved successful, as the implementation of the tax was made conditional on other OECD countries following suit, which was not going to happen. Furthermore, the United Kingdom rejected it decisively in 1993, and other member states did not come out with strong support for the proposal either. This meant the de facto burial of the carbon tax. In both cases, U.S. and EU businesses proved to have veto power over the choice of policy instrument in climate politics. Carbon taxes were a no-go in the eyes of industry, as they were perceived to impose prohibitive compliance costs. The experience with the defeat of carbon/energy tax proposals will likely have a lasting effect on policymakers on both sides of the Atlantic. They acknowledged that some form of business support for the choice of instrument was crucial in order to be able to pass mandatory climate policy, as will be shown later in this and the following chapters.

The fact that the UNFCCC did not include targets and timetables, and the defeat of the energy/carbon tax proposals in the United States and the EU, represent victories for the antiregulatory business coalition in the face of opposition from the environmental movement. The success of the GCC and its allies relied primarily on superior financial resources, the GCC's broad membership and associated legitimacy, powerful state allies such as the U.S. administration, Congress, and the Organization of Petroleum Exporting Countries (OPEC), and the ability to organize opposition among a number of key powers and at the international level.

Yet on the international scene, COP 1 in Berlin in March–April 1995 marked a turning point in the negotiations. The task at COP 1 was mainly to review the UNFCCC's commitments and decide on the next steps. While in the run-up to the conference governments could not agree on further necessary steps, the conference itself unexpectedly emerged as a watershed in international climate policy through the agreement of the so-called Berlin Mandate. The mandate said that governments should "set quantified limitation and reduction objectives within specified timeframes" by 1997 (United Nations 1995, 5). This leap was mainly the result of a split in the G-77 representing developing countries. A so-called Green Group of developing countries led by India emerged and coalesced with the EU, calling on developed countries to take on binding emissions reduction targets (Oberthür and Ott 1999). The mobilization of a coalition for targets and timetables was due to the efforts of the global environmental movement. Environmental groups had organized effectively in the Climate Action Network (CAN) (Alcock 2008; Betsill 2008a), which successfully pushed for international GHG caps.[4]

Reportedly, oil firms were actively involved in lobbying against quantified targets and timetables in particular at the last preparatory meeting for COP 1 in New York in February 1995 (Grubb, Vrolijk, and Brack 1999). They tightly coordinated strategies with OPEC countries (Leggett 2001). While the GCC lobbied strongly against new commitments, the first intraindustry disagreements emerged at COP 1. Exxon and Texaco continued to pursue an offensive strategy along with U.S. coal interests, whereas BP and Shell started to soften their antiregulatory stance. Most likely this was because European firms became cautious about damaging the image that firms could contribute to solving environmental problems that they had built up in the Rio process (Grubb, Vrolijk, and Brack 1999). While at COP 1 these differences between firms were subtle, they foreshadowed the emergence of a rift in the global business community, with significant impacts on policy development. Why did the antiregulatory coalition begin to lose power even as a new coalition was emerging?

The Industry Split and the Campaign for Market Mechanisms, 1995–1997

In July 1996, governments met in Geneva for COP 2. During COP 1 in Berlin, the United States only agreed to the mandate against mounting pressure from other states, but at COP 2 the United States performed the U-turn that the Berlin Mandate had announced (Bodansky 2001). Heading the U.S. delegation to Geneva, Wirth made a strong statement in support of binding targets and market mechanisms. For the first time in the negotiations, emissions trading was on the table. The new U.S. position was widely seen as a sea change, exciting the environmental community and enraging many businesses. The U.S. demonstration of political will lent strong support to the advocates of a binding international treaty, leaving opponents such as the OPEC countries and Russia alone.

Against this backdrop, COP 2 produced the Geneva Declaration, which reaffirmed the Berlin Mandate, and made it clear that the agreement would not only be legally binding but would also include targets and timetables. The shift in U.S. position toward quantified targets and timetables was inextricably linked to flexible mechanisms. At the time, the U.S. administration itself was unclear about how emissions trading could be realized in the context of an international agreement (Grubb, Vrolijk, and Brack 1999). The application of the instrument in an international context only became more apparent when a group of U.S. officials retreated over Christmas 1996 to develop text for a protocol (interview 42). This import of a new instrument into the negotiating process had been preceded by a struggle between industry factions over the course of U.S. foreign policy on climate change.

Business Conflict and the Emergence of the Protrading Coalition

"Cracks appear in the carbon club," explains Jeremy Leggett (2001, 245), summarizing his observations of business advocacy at COP 2. At the time, the GCC was still a powerful actor, and continued to be so until the Kyoto conference and beyond. A split over political strategy was emerging in industry, though, in particular between European and U.S. oil companies, which were each facing different regulatory contexts at home (Levy and Kolk 2002). While business conflict started to emerge as early as 1995–1996, it only led to an outright split of industry in spring 1997.[5] The companies defecting from the mainstream industry stance formed a loose NGO-business coalition centered around market-based climate policy. This represented a distinct shift in the political landscape: the first phase of climate politics had been a battle between a relatively monolithic bloc of firms against a strong environmental advocacy network (Pulver 2004). In the proemissions trading coalition, however, a number of firms and U.S.-based environmental nongovernmental organizations (ENGOs) found common ground by promoting emissions trading as the compromise solution. This sidelined firms that rejected mandates, such as the GCC, and ENGOs that opposed market mechanisms but preferred domestic policies and measures, such as the Sierra Club. It is important to note that the companies that left the GCC did not pursue individual political strategies but instead coordinated their activities with other firms and NGOs. Hence, the conflict in the business community became only politically salient when a new, somewhat-organized force emerged.

Four organizations were at the core of the early protrading coalition: Environmental Defense, the International Climate Change Partnership (ICCP), BP, and DuPont. Environmental Defense, the "apostle of emissions trading," as a policymaker says, had been a longtime advocate of emissions trading for GHGs (interview 46; Samson 1998). The idea of SO_2 trading had been pioneered by Environmental Defense economists in view of the application of emissions trading in climate policy (interview 24; Joskow and Schmalensee 1998). The positive historical experience with SO_2 trading under the Acid Rain Program convinced Environmental Defense to also support emissions trading in the case of climate change. The actual expenses of compliance with the SO_2 program were at 30 percent of the projected cost due to an efficient trading scheme (Chameides and

Oppenheimer 2007). The tensions within the oil industry provided Environmental Defense with an opportunity to mobilize industry support for its longtime political project of emissions trading. Other NGOs that were supportive of emissions trading included the World Resources Institute (WRI) and the Natural Resources Defense Council (NRDC), though they did not take such a public profile as Environmental Defense.

The latter intentionally marketed itself as a "business-friendly" NGO. Fred Krupp, president of Environmental Defense, said at Kyoto: "The US needs trading. Once you create a profit motive for enterprises to invest you unleash forces that are desperately needed to solve a problem like global warming" (quoted in Boulton and Clark 1997, 6). With its market-oriented and business-friendly approach, Environmental Defense alienated the core of the environmental community, breaking with CAN, even though it had once been the network's founding member (Pulver 2007; Alcock 2008). CAN members advocated substantive domestic emissions reductions through domestic policies and measures instead of international emissions reductions. Carbon trading was only considered a supplementary policy instrument. Hence, business conflict went along with NGO conflict.

On the industry side, the ICCP, BP, and DuPont became the most vocal advocates for emissions trading. Convened by lobbyist Kevin Fay of the lobbying firm Alcalde and Fay, the ICCP gathered a number of multinationals from across the economy. Several of them had been actively involved in ozone politics through the Alliance for Responsible Atmospheric Policy, an industry coalition that Fay had set up in 1980. Based on their experience with ozone politics, these companies favored a moderate industry voice that engaged with climate policymaking in a constructive way (interview 33). Accordingly, the ICCP's slogan was, in Fay's words, "to make the process smarter, not faster or slower" (interview 33). The ICCP was never a proregulatory coalition in the sense that it called for climate regulation, but it emerged as a keen supporter of emissions trading when international mandates became increasingly likely. The change in political strategy was preceded by internal change when Kathryn Shanks from BP became the ICCP's chair in 1995–1996. This forced BP to develop a clear policy position on climate change, which affected its own highly visible leadership and the ICCP's strategy (interview 33).

The ICCP's advocacy was closely tied to the rise of BP as a climate leader and political entrepreneur. The company quickly became the key player in the NGO-business coalition advocating international carbon trading. When Shanks became chair of the ICCP, BP also began to focus on the climate issue internally. In January 1996, a seminar was organized for John Browne, the CEO of BP, and the executive committee, in which John Houghton, the then chair of the IPCC, gave evidence of global climate change (interview 7). From this early moment on, Browne took personal leadership on the issue with the intent to stand out. According to insiders, a number of reasons explain Browne's leadership: as an engineer he understood the severity of the climate change problem; BP had studied business involvement in ozone politics; BP was a member of the ICCP; and BP hoped to make a case for emissions trading by demonstrating that it works (Victor and House 2006).

BP's leadership on climate change reflected the general role that the company took within the industry under Browne: "He clearly has the first-mover advantage and the industry is dancing to his tune," said Robin West, a U.S. industry expert (quoted in How Green Is Brown? 1999). For instance, BP's merger with Amoco in 1998 set off a round of mergers in the oil industry. Browne himself said that he was convinced that oil companies "must engage in the debate, and not be shut out as the bad guys, I want us to be . . . I want us to be—dare I say it—progressive" (quoted in ibid.). After COP 2, in fall 1996, following a request from BP's leadership to learn more about emissions trading, Environmental Defense and BP signed a memorandum of understanding (interview 24). Thereafter, Environmental Defense created an educational program on emissions trading for BP. In addition, Krupp lobbied Browne to adopt a trading scheme (Victor and House 2006).

When BP teamed up with Environmental Defense, it decided to publicly break with the GCC in October 1996, which made the latent business divide a full-blown business conflict. With this move it followed DuPont, which had quietly left the coalition earlier the same year. BP's breakup with the dominant coalition along its cooperation with Environmental Defense caused an uproar within the oil industry and broader industry alliance (ibid.). The company was perceived to be "leaving the church," as a company representative put it (interview 7). The political leadership of BP would prove to be crucial in reorganizing the coalitional landscape of business interests.

On May 19, 1997, Browne (1997) gave a famous speech at Stanford University, in which he supported emissions trading and announced BP's collaboration with Environmental Defense on an internal trading scheme. By then, the collaboration had matured to an official partnership. The speech was based on an internal "Where BP Stands" document that had been developed in late 1996, and "road tested" with NGOs and governments in early 1997 (Lowe and Harris 1998). Ever since this landmark speech, BP was the figurehead of the NGO-industry campaign for international emissions trading. While the speech was met with the fury of mainstream business, other businesses such as DuPont were pleased (interview 25). DuPont, a key actor in ozone politics and the ICCP, had been a longtime advocate of emissions trading.[6] Its rationale for promoting emissions trading was twofold. From a long-term perspective, the company wanted to ensure cost-effectiveness. In the short term, DuPont was interested in credits for early action because the company had been reducing GHG emissions since the early 1990s (interview 31). It had gained early expertise in emissions trading through the participation of its Canadian branch in Ontario's Pilot Emission Reduction Trading project, which had been launched in 1996 (interview 31).

BP also mobilized support for emissions trading in the European business community. The BP representatives Klaus Kohlhaus and Michael Wriglesworth were the first two chairs of the Working Group on Climate of BusinessEurope, the largest industry association at the European level (interviews 7 and 11).[7] Between February and September 1997, BusinessEurope developed its policy position for COP 3, which endorsed emissions trading as a crucial building bloc of the Kyoto framework.[8] BP's leadership also directly affected Shell's stance on climate change (interview 17). Shell soon adopted a proactive climate strategy, though it only became an advocate for emissions trading after the Kyoto conference. After adopting a sustainability policy in its "Statement of General Business Principles" in March 1997, Shell was the first oil company to acknowledge the need for precautionary action in September 1997 (Oil and Gas Newsletter 1997), although the company did not refer to emissions trading.[9] In fact, in mid-November, Chris Fay, Shell United Kingdom's chief executive, said that the company thought a carbon tax was a good instrument in fighting climate change (Leggett 2001). It was only in February 1998 that Shell started to consider an involvement in the future emissions market (Rank 1998). Henceforth, it became a central player in the NGO-business coalition driving international carbon trading. In April 1998, the company left the GCC (Shell 1998).[10]

Together Environmental Defense, the ICCP, BP, DuPont, and later Shell constituted the heart of the emerging NGO-industry coalition revolving around international emissions trading as the compromise solution. The early movers in the business community were not proactive in the sense that they were pushing for climate regulation. Rather, they were interested in ensuring that emissions trading would be the policy response in case governments decided to implement binding carbon constraints (interview

7). Emissions trading emerged as an acceptable formula for big emitters because it minimized the economic costs of environmental regulation. BP and the other pioneers viewed emissions trading as the most cost-effective option of all available policy instruments. In particular, a carbon tax was seen to be the greater evil compared to emissions trading. Several U.S. firms had reduced their GHG emissions through voluntary programs that the Clinton administration had launched in 1993 (interviews 25 and 34). They had considerable interest in receiving credits for early action, which they could then sell in a trading scheme. Interestingly, financial services companies were not among the early supporters of international carbon trading, though the policy instrument held the potential to create one of the largest financial markets.

Coalition members exchanged knowledge and information, and loosely coordinated their advocacy strategies regarding GHG emissions trading. From the beginning, BP played an outstanding role as the broker and hub of the network. The company made the connection with the environmental community, created support for emissions trading among the U.S. constituency of the ICCP, and mobilized support for emissions trading in the European business community. Hence, BP bridged two divides: the one between business and the environmental community, on the one hand, and that between U.S. and European business, on the other hand. While the firm did not pursue its new political strategy by itself, it was extraordinarily influential in organizing and shaping the protrading NGO-business coalition. Other business groups did not actually belong to the NGO-business coalition but instead quietly supported emissions trading. The Edison Electric Institute (EEI), the representative body of the U.S. utility sector, was a GCC member, yet it strongly supported emissions trading and communicated this to the administration (interviews 38 and 45). The National Association of Manufacturers (NAM) pursued the same strategy. While these businesses opposed mandates in first place, they were in favor of emissions trading if emissions limits could not be prevented.

The Protrading Lobby and U.S. Foreign Policy on Climate Change

The emergence of a split in industry and the rise of a protrading coalition coevolved with the Clinton administration's pushing for a global climate regime based on emissions trading. Business conflict and the subsequent emergence of a protrading lobby figured critically in Clinton's decision to support Kyoto (Levy and Egan 2003). It is not the case that business and NGO advocacy can fully be credited for the change in U.S. foreign policy. Rather, most observers say, it was a reciprocal process, in which the business coalition lobbied the administration, and in turn the administration was actively seeking business support (interviews 46). This provides evidence for the general notion that the rise of carbon trading is due to multilevel leadership from proactive firms, market-oriented government actors, and business-friendly NGOs. Business support for emissions trading created enabling conditions for governments to move forward with mandatory climate policy.

President Clinton had entered his reelection race with an ambitious environmental program. After reelection in 1996, he advocated a stabilization of emissions at 1990 levels by 2008–2012. Market mechanisms seemed to be the natural choice for implementing GHG controls for two reasons beyond the emergence of a protrading NGO-business lobby. For one, since the first Bush administration, subsequent administrations have had an institutional bias toward emissions trading as the preferred implementation tool for pollution abatement policies (interview 42). The success of the domestic trading scheme for SO_2 in the U.S. electricity sector strengthened governmental support for emissions trading. Some say that the administration was unanimous regarding emissions trading, while others point to the EPA's role in internally advocating the instrument (interviews 45 and 30). Second, the Kyoto framework was designed in the mid-1990s, during which the debate on national competitiveness critically affected policymaking. As a key official from the Clinton administration said, "Everything had to be framed in economic terms" (interview 46). Given these reasons, the administration was highly receptive to business interests making a case for emissions trading.

The ICCP, in particular, worked closely with the Clinton administration early in the process. In fact, the Clinton administration sought the ICCP out as the preferred industry voice (interview 46). The ICCP claims to have been—and is partly credited with being—the first business group laying out the market-based framework of the Kyoto Protocol (Fay 2005). During COP 2, the ICCP (1996b) published a press release supporting "essential elements of a policy framework" that includes "a mixture of policy instruments including joint implementation and emissions trading."[11] Again, the legacy of the ozone case was helpful: Fay had a working relationship with Eileen Claussen in the U.S. Department of State that dated back to the Montreal Protocol's negotiations. In the run-up to COP 2, Claussen led the interagency process, which ensured the ICCP good access to policymakers (interview 42).

In February 1997, the White House Climate Change Task Force was set up to develop the administration's climate policy. In June of the same year, President Clinton and Vice President Al Gore launched the White House Initiative on Global Climate Change to create support for strong climate action. Part of this campaign was to mobilize support for market mechanisms. In October 1997, Clinton finally revealed details of his climate change policy at a White House conference (Boulton 1997), asserting that he aimed to "harness the power of the free market" in mitigating GHGs. He continued to say, "If we do it right, protecting the climate will yield not costs, but profits, not burdens but benefits, not sacrifice, but a higher standard of living" (quoted in Warrick and Baker 1997).

In a situation where climate change politics was increasingly caught in a stalemate between business and environmental interests, the protrading NGO-business coalition supported a compromise solution. The cooperation of antagonistic interest groups—that is, big energy companies and environmental groups—behind the same policy idea gave particular weight to the idea of a market-based climate regime. The new coalition offered a solution that would help overcome the policy crisis. This occurred in an international climate in which the scientific and environmental communities were increasingly demanding some form of mandatory climate regulation, thus working toward establishing an international norm for climate action.

The protrading NGO-business alliance and the Clinton administration's foreign policy coevolved in a highly charged political terrain. Though it was no longer the one-and-only voice of industry, the GCC and its allies continued to question the science and economic viability of internationally binding emissions controls (Pulver 2005). In the battle over domestic U.S. climate policy, the network exercised considerable influence in Congress. On June 8, 1997, the Business Roundtable published full-page advertisements in U.S. newspapers (Levy and Egan 2003). Signed by 130 CEOs, these ads argued against the adoption of mandatory emissions reduction targets at the upcoming conference in Kyoto. The U.S. Senate ultimately proved to be the critical channel of influence for industry. On June 12, it agreed on the bipartisan Byrd-Hagel resolution, which states that the Senate would not ratify an international agreement on emissions reduction targets "unless the protocol or other agreement also mandates new specific scheduled commitments to limit or reduce greenhouse gas emission for Developing Country Parties" (U.S. Congress 1997). The struggle between antiregulatory business groups and the protrading coalition continued to play out at the Kyoto conference.

The Kyoto Protocol: The Emissions Trading Compromise, 1997

Influencing U.S. foreign policy in favor of emissions trading was only the first step in the internationalization of emissions trading. Including the emissions instrument in an international treaty represented the second step. The transnational NGO-business coalition had left its mark on U.S. foreign policy, but it was in no way clear whether an actual deal could be struck between the EU and the United States. European ENGOs and governments had been staunchly opposed to emissions trading as a policy instrument. Environmental groups such as CAN initially perceived it as a "license to pollute," and preferred domestic command-and-control regulation, while governments had reservations as to its feasibility. Up until shortly before COP 3, the political process was highly polarized between the EU and developing countries, on the one side, calling for targets and timetables along with policies and measures, and on the other side, the United States and its allies in the JUSCANNZ group objecting to support targets without flexible mechanisms (Bodansky 2001).[12]

The first sign of a compromise that included emissions trading was a shift in the EU's position on the policy instrument in the run-up to COP 3. In June 1997, the European Council of Ministers adopted compromise language under the leadership of the Dutch presidency and with strong support from the UK government: "The council considers that mechanisms such as emissions trading are supplementary to domestic action and common coordinated policies and measures, and that the inclusion of any trading system in the Protocol and the level of the targets to be achieved are interdependent. It therefore calls upon all industrialized countries to indicate the targets they envisage for 2005 and 2010 (cited in Grubb, Vrolijk, and Brack 1999)."[13] Nonetheless, the final session of the Ad Hoc Group on the Berlin Mandate in October 1997 in Bonn ended in a deadlock. Kyoto had to deliver it all.

Business Coalitions Competing for Influence at Kyoto

Against this background, delegates—both governmental and nongovernmental—convened in Kyoto from December 1 to 11, 1997. Altogether around four thousand nongovernmental delegates attended the conference (Depledge 2005, 211). Headed by Gail McDonald, the GCC came with sixty-three delegates to the conference, including representatives from the API, the American Automobile Manufacturers Association, and the U.S. National Coal Association (Leggett 2001, 290). The GCC and its allies tried to prevent the U.S. government from agreeing to obligatory reduction targets in the first place.

Since the proregulatory alliance did not have the same level of organization, it is difficult to establish how many lobbied for its interests. Key actors, again, were Environmental Defense, BP, the ICCP, the World

Business Council for Sustainable Development (WBCSD), and the Business Council for Sustainable Energy (BCSE).[14] They played a pivotal role in mobilizing support among governments for emissions trading, which was the sine qua non for the United States to agree to targets and timetables.[15] They assumed the role of knowledge providers and brokers in the international negotiations. In particular, Environmental Defense and BP briefed delegates about emissions trading as an instrument (interviews 7 and 24). Charles Nicholson, from BP as well as chair of the WBCSD's Climate Change and Energy Working Group, had good access to climate negotiators and could educate them about emissions trading. BP, in addition, was central in convincing the EU to accept the U.S. proposal on emissions trading by lobbying John Prescott, the environment minister in the newly established government of Tony Blair (interview 45). Prescott in turn persuaded other European member states to accept the deal (interview 39). Finally, BP organized business support for Kyoto in Europe. Speaking for BusinessEurope, BP manager Kohlhaus said: "UNICE and industry in Europe are prepared to support targets" (quoted in Leggett 2001, 310).

Next to the protrading lobby, other progressive business forces supported Kyoto in general, while not especially lobbying for the inclusion of flexible mechanisms. These forces involved firms whose products would be in higher demand under any kind of emissions control, including the renewable energy sector, the nuclear industry, the natural gas industry, and others. The European and U.S. branches of the BCSE primarily represented these companies, though some of them also formed the ad hoc alliance Business Challenge (Leggett 2001). On the first day of the Kyoto negotiations, Business Challenge (1997) published a full-page ad in the *Wall Street Journal*, calling on "the United States government to exert strong leadership by promoting climate change policies that provide incentives to act quickly." A key company in both Business Challenge and the U.S. BCSE was Enron, which vocally supported Kyoto in the hope of increasing demand for natural gas (Yandle and Buck 2002). Hence, for Enron, climate policy provided a win-win set: "We all know with confidence that appropriate steps to respond to climate change—based upon the efficient and clean use of energy—will lead to long-term, worldwide economic growth," said Mike Marvin, the U.S. executive director of the BCSE, in an intervention at COP 3 (quoted in Leggett 2001, 311).

The GCC took the exact opposite stance by trying to prevent agreement on targets and timetables. It closely coalesced with Saudi Arabia and other oil-exporting countries in stalling negotiations. John Grasser, of the U.S. National Mining Associations and a GCC representative, was outspoken: "What we are doing, and we think successfully, is buying time for our industries by holding up these talks" (quoted in ibid., 301). While the U.S. administration was not an ally of the GCC, the U.S. Senate was. Senator Chuck Hagel of Nebraska, a particularly vocal supporter of fossil fuel interests, stated: "Entire industries will leave the United States for countries that won't be bound by this treaty. For the first time in American history we would be giving an international body the authority to limit and regulate our economic growth" (quoted in ibid., 309).

With the United States moving toward acceptance of binding targets as long as emissions trading was part of the package, the GCC grew increasingly frustrated. On the occasion of Gore's speech in Kyoto, the GCC (1997) issued a press release saying: "The only new markets that will be created by additional government regulation will be markets of inefficiency; companies straining under ill-conceived political targets that have no connection to real need or sustainable environmental progress. This will not bring about the most cost-effective policies, and it certainly will not help American workers whose jobs are at stake in Kyoto."

Governments Striking a Compromise Deal

While business groups and NGOs worked to influence the different coalitions at Kyoto, and tried to bridge the divide, governments struck the final deal. The United States had come to Kyoto signaling that it would not agree to an international agreement at any cost. In the first week of the negotiations, the United States did not move toward a strong emissions reduction target. Kyoto was at risk of ending in a fiasco. The tide turned when U.S. vice president Gore visited the conference on December 8 to give a speech.[16] Reiterating the U.S. position on market mechanisms and the participation of developing countries, Gore announced greater U.S. flexibility in working toward an agreement (International Institute for Sustainable Development 1997). This lent negotiators the flexibility and bona fides to negotiate a compromise (interview 45).

From then on, the negotiations made progress, though the issue of emissions trading was not resolved until the final plenary session. The United States had formed the so-called Umbrella Group[17] to counterbalance the EU's position on the level of emissions reductions and trading. In this situation, the United States stressed its willingness to enter strong emissions reduction commitments if flexibility mechanisms were included. Stuart Eizenstat, head of the U.S. delegation, said: "I urge in the strongest terms, now that these historic commitments are about to be made, not to deprive us of the means to meet them. The eyes of the world are upon us. We have the opportunity to do something great" (quoted in Leggett 2001, 317). He continued that market mechanisms were absolutely necessary to do that.

To resolve the stalemate, the chair, Argentinean ambassador Raúl Estrada-Oyuela, unilaterally drafted a new article that suggested that the modalities and rules of emissions trading should be dealt with at COP 4, as an agreement could not be reached at COP 3 (Boulton and Hutton 1997). Estrada-Oyuela said that the COP should consider the conditions for "the new animal" before letting it "run wild in different places" (quoted in International Institute for Sustainable Development 1997, 12). The final article 17 on emissions trading reads as follows:

> The Conference of the Parties shall define the relevant principles, modalities, rules and guidelines, in particular for verification, reporting and accountability for emissions trading. The Parties included in Annex B may participate in emissions trading for the purposes of fulfilling their commitments under Article 3 of this protocol. Any such trading shall be supplemental to domestic actions for the purpose of meeting quantified emission limitation and reduction commitments under that Article. (cited in UN 1998, 12)

Emissions trading became a central pillar of the Kyoto Protocol as the result of a deal between the EU and the United States, largely against the opposition of developing countries. This is noteworthy, since developing countries would later emerge as the major beneficiaries of international emissions trading. Yet in Kyoto, India, speaking on behalf of the G-77, and China objected to emissions trading, arguing that it would delay and undermine the domestic emissions reductions of industrialized nations. Instead, the G-77 and China pushed for ambitious emissions reduction targets for developed countries, while opposing an article that called for voluntary emissions reductions from developing countries (International Institute for Sustainable Development 1997).

A notable exception is the case of the CDM. In June 1997, Brazil had proposed a "clean development fund," which was originally intended as a vehicle to finance projects in developing countries through penalty fees paid by annex I countries for noncompliance (Streck 2004; Lecocq and Ambrosi 2007). The idea was that if an annex I party failed to meet its emissions reduction target by the end of a budget period, it would have to pay into the fund. The G-77 and China later endorsed the proposal, while industrialized countries opposed penalties for noncompliance. Yet the United States liked the flexibility that the instrument seemed to offer to annex I parties that struggled meeting their targets. U.S. delegates interpreted the Brazilian proposal as a "trading system" and "flexible financing instrument" (Werksman 1998, 151–152). In Kyoto, Brazil and the United

States worked together to develop the CDM as a market mechanism. The two countries managed to mobilize other supporters, including Mexico and the Democratic People's Republic of Korea. According to an observer, the supporters of the CDM from developing countries "had their national interests explained to them" (quoted in International Institute for Sustainable Development 1997, 15).

In Kyoto, emissions trading had found its way into an international agreement, although many of the details remained to be clarified. Gore concluded, "In the end, the final agreement was based on the core elements of the American proposal" (quoted in Kyoto Aftermath 1997). He added that freer markets, not governments, would address the problem of climate change. A small group of protrading advocates from NGOs, business, and within the administration had thus helped to internationalize a U.S.-specific regulatory approach. The future of emissions trading, however, largely depended on the ratification process and domestic politics. The protrading coalition had placed emissions trading on the agenda of international climate politics, but the antiregulatory coalition determined the course of domestic climate politics.

Institutionalizing the Protrading Lobby, 1998–2000

The Kyoto Protocol agreement provided the emissions trading approach with considerable legitimacy. It could, though, by no means be taken for granted that international GHG emissions trading would become a success story. First, there was considerable uncertainty as to what emissions trading actually meant. It was more a formula for cost-effective abatement than a clear-cut policy instrument. Second, the devil lay in the details: the rules for the use of flexible mechanisms had not yet been spelled out—and negotiators had deferred a number of open issues to the next COP. Third, it was not clear that the Kyoto Protocol would ever be ratified by the United States, given opposition in the U.S. Congress. Yet without U.S. participation, neither the protocol nor international emissions trading were likely to fly. And finally, strong aversions against the use of trading mechanisms existed across governments, business, and the environmental community within the EU, the strongest supporter of the protocol.

The NGO-business coalition promoting market-based climate policy was critically aware of the political opportunities and threats in the aftermath of the Kyoto conference. In the attempt to increase its political clout, the coalition created new organizations, thus entering a phase of institutionalization that would enhance its organizational capacity. Overall, the protocol had shifted the discursive balance of power from the antiregulatory camp to proregulatory business groups. While the protrading coalition intensified the level of transnational cooperation, the GCC's disintegration continued.

New Platforms for the Anglo-American Protrading Lobby

While the main advocates were only loosely organized before Kyoto, they now launched new cooperative initiatives between NGOs and business as well as organizations that provided platforms for advocacy, information sharing, and rule setting. In particular, U.S. ENGOs joined forces with businesses from both sides of the Atlantic. While some of the new organizations were clearly endorsing the Kyoto Protocol, others were more cautious in their positioning on the treaty, though all of them were in favor of emissions trading. Key initiatives included the continued cooperation of BP and Environmental Defense on BP's internal trading scheme, the Pew Center's Business Environmental Leadership Council (BELC), the WBCSD's GHG Protocol Initiative, and most important, the IETA.

BP's collaboration with Environmental Defense in the run-up to the Kyoto conference represented the heart of the NGO-business coalition. It had built a bridge between the formerly antagonistic camps of environmentalists and corporate executives, clearing ground for the emissions trading compromise. After Kyoto, the collaboration resulted in the development of BP's internal emissions trading scheme, which was launched in September 1998 (Victor and House 2006). The trading scheme was bound to a voluntary commitment to a 10 percent reduction target by 2010 compared to 1990 levels. BP's trading scheme was the first real-world test, which demonstrated that GHG emissions trading could work at a global scale. Indeed, the main purpose of the scheme was political. BP aspired to show that emissions trading was a viable and effective policy instrument to forestall more costly policy responses, such as a carbon tax or the command-and-control policies that were prominent in Europe at the time. Henceforth, BP's pioneering work became the reference point for public and private actors that developed emissions trading schemes. The collaboration between BP and Environmental Defense marks the beginning of a phase of institutionalized engagement between industry and NGOs on climate change and emissions trading. The Pew Center closely followed in the footsteps of these early pioneers.

In April 1998, the Pew Center on Global Climate Change was established in Washington, DC. The think tank was the brainchild of Eileen Claussen, a former U.S. assistant secretary of state for oceans and international environmental and scientific affairs, who had resigned from government service in mid-1997.[18] Claussen was critically aware of the role of business in environmental negotiations through her involvement in the negotiations leading up to both the Montreal and Kyoto protocols. Based on this experience, she set up the BELC as an initiative of the Pew Center. The BELC initially attracted thirteen companies that took a more progressive stance on climate change. Member companies endorsed a set of principles, but they did not use the BELC as an advocacy platform. It was rather an information-sharing group (interview 20). As an industry representative said, the BELC's principles "danced a fine line on Kyoto" (Changing the Climate of Opinion 2000). While cautious in endorsing Kyoto, the BELC stood firmly behind market-based mechanisms: "The Kyoto agreement represents a first step in the international process, but more must be done both to implement the market-based mechanisms that were adopted in principle in Kyoto and to more fully involve the rest of the world in the solution" (Pew Center 1999).

Early members of the BELC included BP, but also American Electric Power, a heavily coal-based electric utility company. The Pew Center became a high-profile platform for market-based climate policy, and Claussen emerged as the figurehead of the campaign for international GHG emissions trading, not least because she spent large amounts on advertisements. Claussen herself said: "In the past, only the bad guys were out there in the media. We raised the ante by being so public" (quoted in Changing the Climate of Opinion 2000). The following memo from Clinton to Blair offers proof of the Pew Center's rise to fame along with the fact that the new NGO-business coalition was being listened to at the very top:

> Dear Tony, Knowing our shared interest in global climate change, I have enclosed a report on the benefits of greenhouse gas emissions trading. . . . This new report by the non-profit Pew Center on Global Climate Change supports the view that a flexible, well-designed trading system will significantly reduce the costs of climate change mitigation. . . . With best regards, Bill. (cited in ibid.)

Frank Loy, a top U.S. climate change negotiator at the time, said that corporations had "shifted from being climate sceptics to climate activists" (quoted in ibid.), attributing a large part of this change to Claussen.

One of the first activities of the BELC was a joint report with the Pew Center on credits for early emissions reduction efforts of companies. In concert with several business groups and ENGOs, the Pew Center aimed at initiating legislation in Congress that guaranteed corporate climate leaders credits for early action in Congress. The problem, however, was that ensuring credits for early action was tantamount to issuing a guarantee for some form of mandatory climate policy in the future. Congress was far from willing to do this, which is why the initiative failed (interview

33). Next to the BELC, the WBCSD represented a forum for companies supporting a market-based global climate regime.

Based in Europe, the WBCSD was an incubator for the corporate sustainability agenda throughout the 1990s—having been created in the run-up to the Rio Earth Summit (Bernstein 2001). The organization was involved in lobbying for flexible mechanisms at the Kyoto conference and promoted the protrading agenda in the post-Kyoto phase mainly through two initiatives: the GHG Protocol Initiative, and the creation of the IETA. Jointly run by the WBCSD and the WRI, the GHG Protocol Initiative is a multistakeholder initiative that developed an international accounting standard for GHG emissions reductions.[19] The stakeholder process was begun in 1998, and the protocol was launched in October 2001. The GHG protocol is not necessarily linked to the emissions trading agenda, as any kind of voluntary or mandatory emissions reduction program requires accounting standards and GHG inventories. Nevertheless, the rationale behind the initiative was very much to create a standard for corporate emissions reductions endorsed by ENGOs and governments to ensure credits for early action under a future mandatory scheme (interview 4). For instance, DuPont, an early mover in terms of emissions reductions, had a strong interest in the standard.

The most significant organization emerging in the years directly following the Kyoto conference was IETA, the trade association of the emerging carbon industry. It was created at COP 4 in November 1998 by representatives of the WBCSD, a couple of member companies, and key advocates of emissions trading, such as Frank Joshua of UNCTAD and Richard Sandor from the Chicago-based Centre Financial Products Limited.[20] Its roots date back to June 1997, when the first Policy Forum on Greenhouse Gas Emissions Trading was organized by UNCTAD and the Earth Council in Chicago (Sandor 2003). UNCTAD had already sponsored the first report on emissions trading in climate politics in the run-up to UNCED, as mentioned in the previous chapter.[21] The Earth Council had been cofounded by Maurice Strong, UNCED's secretary general and a crucial advocate for business-friendly solutions to environmental problems. In the run-up to COP 4, it became increasingly evident that an organization was needed that promoted the creation of an international carbon market, though participants were not clear about whether it should represent both business and UN agencies, or be purely private in nature (interview 51). At a meeting during COP 4, the participants finally asked the WBCSD to set up an organization that would foster the cause of emissions trading.

Thereafter, it lay in the hands of Bjoern Stigson (the president of the WBCSD), Joshua, Paulo Protasio (the president of the Latin American Trading Association), and senior executives from BP and Shell to set up IETA in 1999 (Gallon 1999). The association was officially launched at the International Petroleum Exchange in London on June 4, 1999. IETA was incorporated in Geneva and initially housed by the WBCSD. Under the leadership of first Protasio and then Andrei Marcu, IETA rose to become a "broad church," representing both emitters, such as the energy producers and energy-intensive manufacturing industries, and financial intermediaries, such as banks and brokers.[22] Membership in IETA soon became "the rite of passage" into the carbon industry, as a prominent actor said (interview 51). It thus eclipsed the Environmental Markets Association, which previously had been a forum for emissions trading experts.[23] IETA was the most vocal business advocacy group on emissions trading on the international stage. While it first focused on the international process, it later came to play a role in both European and U.S. climate politics. Moreover, the association had considerable technical expertise, which it shared with governments.

In sum, following the Kyoto Protocol agreement, the coalition advocating emissions trading as the climate compromise became an institution. As an influential lobbyist put it, a "relatively closed fraternity" had emerged (interview 33). Its strongholds were businesses from the United Kingdom, the United States, and business-friendly NGOs or think tanks from the United States. Given its key actors, it was very much an Anglo-Saxon advocacy coalition. The European environmental community had yet to be convinced of the value of international emissions trading, which was widely perceived as a cheap buyout of domestic emissions reductions. To this end, Environmental Defense and the Pew Center organized Transatlantic Dialogues on Market Mechanisms in 1998, in which representatives from European ENGOs, business, and governments participated (Petsonk, Dudek, and Goffman 1998).

Through the creation of new organizations, the protrading coalition significantly increased its organizational capacity for information sharing and advocacy work. The creation of IETA especially constituted a strong lobby for international carbon trading that went beyond advocating cap and trade by also creating institutional infrastructure through the promotion of contracts and standards. Moreover, much of the transnational coordination of political strategies occurred in IETA's working groups. The architecture of the protrading network was such that it could target the international process as well as domestic political processes in the EU and North America. Given the continuing struggles over climate policy in both the United States and the EU, it was essential for the coalition to be able to affect change in several countries in order to firmly embed carbon trading in the global climate agenda. Transnational organization created the capacity to affect change given a multilevel governance system with dispersed formal political authority. Also in the phase of the network's institutionalization, BP continued to play a strong role as an initiator of new organizations as well as a broker between countries, business, and NGOs. This suggests that transnational collective action can hinge on only a small number of policy entrepreneurs and brokers.

The Fall of the GCC

As the protrading coalition gained momentum, the GCC's decline continued unabated (Levy and Egan 2003; Gelbspan 2004). By the late 1990s, membership in the GCC was considered a reputational risk (Jones and Levy 2007), which indicates a normative shift toward climate regulation. The outright rejection of carbon controls was slowly but steadily being considered illegitimate by policymakers. BP quietly split from the GCC in October 1996, and Shell publicly announced it would leave the GCC in April 1998. They were followed by a number of U.S. companies, including Dow Chemical, American Electric Power, Ford, DaimlerChrysler, GM, and Texaco (Biello 2003). On leaving the GCC, a Ford spokesperson said, "There is enough evidence that something is happening that we ought to look at this seriously" (quoted in Layzer 2007). Beginning in 2000, the GCC represented only trade associations instead of individual companies. It was dissolved in 2002. In retrospect, William O'Keefe, chair of the libertarian think tank the Marshall Institute and former GCC head, said:

> Prior to Kyoto, the business community was focused, united and forceful. . . . Our support for S.R. 98 [Byrd-Hagel] and our willingness to openly debate the climate change issue and undertake a major advertising campaign was a significant achievement. Momentum was on our side. Unfortunately, we did not sustain it because many in the business community do not understand perseverance. . . . Because of this failure, the [Clinton] Administration and its environmental allies have regained the momentum. They have succeeded in fracturing our unity and putting industry on the defensive. (quoted in ibid, 112)[24]

In short, after Kyoto the protrading NGO-business coalition was gaining political strength, while the unity of its counterplayers disintegrated. This can be described as a changing of the guards between the two forces in terms of organizational capacity and cohesiveness. Though on the decline, the GCC had a last major advocacy success in convincing the Bush administration to withdraw from the Kyoto Protocol, as will be discussed in chapter 6.

Conclusion

While the antiregulatory coalition had a firm grip on U.S. and international climate policy in the early phase of climate politics, business conflict erupted in 1995. Subsequently, business-friendly NGOs and proactively oriented companies joined forces to shape a climate compromise that revolved around binding emissions reduction targets that could be achieved through international emissions trading. Key actors included firms from the oil, power, and energy-intensive manufacturing industries in the United Kingdom and the United States as well as market-oriented NGOs from the United States. The group was essentially an Anglo-American advocacy coalition, which supported carbon trading at multiple political levels, shared information, and generated market infrastructure.

The rise of the protrading coalition marked a significant shift from the topography of political conflict in the early phase of climate politics, which was dominated by conflict between business and NGOs. The new coalition sidelined the more radical antiregulatory business coalition (the GCC) and an NGO coalition (CAN) that preferred domestic policies and measures. The influence of the protrading coalition is best understood as legitimizing the Clinton administration's foreign policy on climate change. For the first time in climate change politics, a number of major corporations supported a mandatory climate policy. Business thus created a window of opportunity for governments to move forward with the mandatory regulation of carbon emissions. It is unlikely that the governments of industrialized countries would have agreed on internationally binding targets without any business support. Business had previously successfully vetoed international emissions reduction targets and carbon-energy taxes in both the EU and the United States (Levy and Egan 1998, 2003; Meckling 2008).

A number of reasons explain the influence of the protrading coalition in shaping U.S. foreign policy in the run-up to Kyoto (table 4.1). The coalition offered a compromise solution in a policy crisis. The polarization of international climate negotiations between the EU and the United States, and between business and environmental interests, had led to a stalemate, which offered opportunities for new compromise solutions. BP, Environmental Defense, and their allies exploited this opening. The protrading advocates offered a third way beyond the options of command-and-control policies, on the one hand, or no emissions controls on the

Table 4.1
The Kyoto Protocol: The sources of influence of the protrading coalition

Source of influence	Impact	Evidence
Political opportunities		
Policy crisis	High	Political polarization between the U.S. and the EU, and between economic and environmental interests
Norms	Moderate	Actors in liberal market economies (the United States and the United Kingdom) were quickly convinced of market mechanisms
Coalition resources		
Financial resources	Low	Protrading coalition had much less financial resources than the GCC
Legitimacy	High	Legitimacy was the key resource in protrading advocacy; the comparative legitimacy advantage stemmed particularly from the fact that two antagonistic actors (businesses and NGOs) cooperated
Coalition strategies		
Mobilizing state ally	High	The U.S. administration and the UK government supported emissions trading in international negotiations; convergence of interests of key powers and the protrading coalition
Playing multilevel games	High	Coalition targeted domestic processes in the EU and the United States as well as the international process
Other		
Institutional memory	High	The U.S. administration had an institutional preference for emissions trading due to its positive experience with SO_2 trading

other. Furthermore, the normative environment for business engagement with climate politics shifted over the course of the 1990s. The advocacy campaigns of environmental groups and more general public pressure in Europe led to a norm of international climate action. The protrading coalition strategically accommodated this pressure.

Interestingly, prior to Kyoto, the protrading coalition did not share financial resources and consisted of only a few organizations, while the GCC was still well funded.[25] This difference in resource equipment does not seem to have translated into the degree of influence of the two groups. The more powerful resource of the protrading coalition was its legitimacy. The fact that two previously opposed interest groups supported a policy lent the coalition the high degree of legitimacy that baptist-and-bootlegger coalitions often hold in policymakers' eyes. Such coalitions offer "prenegotiated" deals between usually antagonistic interest groups. Given the political stalemate dominating climate politics in the mid-1990s, policymakers were willing to listen to a group that offered a way out of the crisis. The case suggests a reciprocal relationship between firms and governments. While business support for a market-based climate policy helped to legitimize the climate policy of the Clinton administration, the protrading coalition could draw on the power of a key state ally.

Given the opposition of the EU and developing countries to market mechanisms, it was crucial for the protrading coalition to be able to ally with the United States, the most powerful player in climate politics. The administration already had an institutional preference for emissions trading based on its institutional memory of the successful SO_2-trading scheme. The coalition thus accelerated a preexisting momentum. Furthermore, in order to shift the global climate agenda toward market-based policy, the campaign had to target a number of major negotiating powers. The coalition tried to make the case for market-based climate policy vis-à-vis policymakers in the United States and EU member states. It is hard to imagine how a single company could have organized an effective transnational campaign given the fact that a number of political processes had to be targeted. Instead, the protruding coalition's pattern of advocacy was that firms communicated among themselves transnationally, while each in turn lobbied those political actors they had good relations with, such as their home governments. It is exactly this duality of global presence and coordination, on the one hand, and local political actions, on the other, that resulted in an effective campaign within the multilevel system of climate governance.

While the GCC largely lost the battle over the Kyoto Protocol at the international level, it achieved the passage of the Byrd-Hagel resolution domestically. The GCC's long-standing relations with the Senate were a critical source of influence. Despite this success, the antiregulatory coalition showed signs of decline. Reasons for the loss of influence over the administration and foreign policy include the accelerating normative momentum for international emissions controls. The GCC's message was resonating increasingly less with the international political discourse. The GCC thus suffered from a decline of legitimacy. Moreover, the GCC lost a major ally in the battle against carbon controls when the Clinton administration decided to support international targets.

In sum, key factors in this transition from one dominant business coalition to another one include the normative shift toward international climate action, the higher degree of legitimacy of the NGO-business coalition, the political crisis evoked by stalemate between business and environmental interests, and the emergence of the Clinton administration as a powerful state ally for mandatory yet market-based climate policy. While Kyoto internationalized what until then had been only a U.S. regulatory instrument, the viability of the emissions trading agenda remained to be tested: Would any signatory of the Kyoto Protocol actually implement emissions trading? Antiregulatory business forces still existed, although they had adopted a less adversarial style and had lost their unity. While after Kyoto, the United States did not move forward on climate policy for a few years, the EU surprisingly came to be the first arena for political action on real-world carbon trading schemes.

5

The European Union: From Foe to Friend of Carbon Trading

Governments, firms, and environmental groups in the EU were the leading skeptics of international greenhouse gas (GHG) emissions trading before Kyoto, and some time after it. The instrument was publicly perceived to be granting a license to pollute to industry and allowing industrialized countries to escape domestic emissions reductions. Yet in January 2005, the EU launched the EU ETS, the first cross-border trading scheme for GHG emissions permits. The scheme covers almost 11,500 entities, which amounts to 45 percent of the total CO_2 emissions in the EU (Egenhofer 2007). When the scheme was first proposed, then EU commissioner for the environment Margot Wallstrom said, "The Proposal on emissions trading represents a major innovation for environmental policy in Europe. We are de facto creating a big new market, and we are determined to use market forces to achieve our climate objectives in the most cost-conscious way" (quoted in Christiansen and Wettestad 2003). Ever since, the EU has been a leader and vocal advocate on emissions trading, and the EU ETS became the yardstick for newly emerging GHG emissions trading schemes around the globe. The EU's U-turn from skeptic to front-runner on emissions trading came as a surprise. The EU was neither forced by other states to adopt the policy nor was it obliged by an international treaty to implement emissions trading. The Kyoto Protocol allowed for emissions trading, but it did not prescribe it.

In this chapter, I explore the process leading to the decision to establish a European carbon market, arguing that UK business and the European Commission led the campaign, which resulted in the import of emissions trading to Europe. I proceed in four steps. The subsequent section shows how firms from the UK oil and power industries pioneered emissions trading, resulting in the emergence of the UK ETS. The UK ETS was the main precursor of a Europewide trading scheme, critically creating momentum for a European scheme. Thereafter, the chapter looks at how a

European protrading business coalition of oil and power companies along with the European Commission moved emissions trading to the top of the agenda in European climate politics. This is followed by an examination of the opposition to emissions trading from Germany and energy-intensive manufacturing industries, and their failure to prevent agreement on a trading scheme. In a fourth step, I demonstrate how the coalition promoting market-based climate policy is at risk of falling apart, due to the existence of the EU ETS. Throughout the narrative, I will analyze why the protrading coalition was influential in the battle over the course of European climate policy. The spotlight will be on the respective roles of funding, legitimacy, state allies, and multilevel strategies.

Pioneering Trading from Bottom-up: Business in the United Kingdom, 1998–2000

The EU ETS was preceded by domestic pilot schemes in EU member states: the Danish and UK schemes.[1] The Danish scheme operated on a small scale, only covering electricity generation (Pedersen 2001; Bode 2005). It did not have a significant impact on the process of developing a European trading scheme (Zapfel and Vainio 2002). The UK ETS, however, was an influential precursor of the EU ETS in at least two ways.[2] The establishment of a national trading scheme created pressure at the EU level "from below" to adopt a Europewide climate policy. In addition, the United Kingdom acted as Europe's backdoor for emissions trading since it was more inclined to adopt market-based environmental policies than the majority of EU member states in continental Europe. The UK ETS was developed between 1999 and 2002, and was operational from April 2002 until March 2007. The trading scheme's creation was the result of a concerted effort of UK industry, notably the oil and electricity sectors, and the UK government to advance emissions trading in Europe. The following sections examine the political effects of in-house trading of corporate leaders and the role of the UK Emissions Trading Group (ETG), an industry organization, in the development of the UK ETS.

Oil Majors as Policy Entrepreneurs: BP's and Shell's In-house Trading

BP's influence on the take-up of emissions trading in Europe was considerable. The company endorsed the Kyoto Protocol, put its stamp on the policy positions of several business associations, and pioneered GHG emissions trading through an in-house trading scheme. In the late 1990s, the company had a clear message to policymakers: "BP Amoco remains convinced that trading has considerable potential to reduce greenhouse gas emissions at least economic cost" (Grice 1999). BP, in short, continued to be the key broker and hub of the protrading coalition. Its protrading campaign benefited from the normative and institutional backing of the Kyoto Protocol, which had created an international momentum for emissions trading. Bernstein and Benjamin Cashore (2000, 80) have pointed out how international rules create a "pull toward compliance," which can be strategically levered by transnational actors: "The rule also becomes a resource on which transnational and/or coalitions of domestic actors can draw when governments do not comply."

The reasons for BP's extraordinarily vocal support for emissions trading are threefold, as Victor and House (2006) suggest. First, the company was seeking experience with the policy instrument that was likely to become a major building bloc of climate policy. Second, BP aspired to demonstrate that emissions trading was a viable and effective policy instrument to forestall more costly policy responses, such as a carbon tax or the command-and-control policies that were prominent in Europe at the time. Third, as a decentralized mechanism, emissions trading allows business units to identify those mitigation options with the lowest marginal cost. BP thus was attracted to the mechanism for the same reason that has attracted governments to decentralized incentive-based market mechanisms, as opposed to command-and-control instruments.

BP's endorsement of emissions trading is closely tied to its engagement with Environmental Defense (Langrock 2006). The company partnered with the environmental organization both to learn about emissions trading and develop an internal emissions trading scheme. At the same time, Browne committed BP to a 10 percent reduction target by 2010 compared to 1990s levels.[3] Browne arrived at this target after BP managers had estimated that 6 to 7 percent reductions could be achieved without incurring costs (Reinhardt 2001). Further consultations suggested that those estimates were likely to be exceeded. Initially the scheme covered 12 business units, but it was expanded to involve all 150 of the company's business units in January 2000.

As BP had anticipated in its strategic reasoning that led to an in-house trading project, emissions trading in Europe proved to be a major political push (Zapfel and Vainio 2002). As the first real-world test of GHG emissions trading, BP's scheme produced facts that supported the cost-effectiveness of the instrument. This lent BP's advocacy efforts strong credibility. Henceforth, BP's pioneering work became the reference point for public and private emissions trading schemes. While leading the industry pack, BP was cautious not to lose touch with industry peers. Rodney Chase, BP's deputy chief executive said in 1999 that being perceived as the green leader was "full of hubris." He added, "I don't think it would be a good thing to set a goal of getting so far ahead of the rest of industry that you have no influence over events" (quoted in Houlder 1999b, 4).

In January 2000, Shell launched its own scheme, the Shell Tradable Emissions Permit System, following the lead of its competitor. Again, Environmental Defense had lent a helping hand in setting up the scheme (Schrope 2001). Shell's permit system had been designed to be compatible with the Kyoto Protocol's provisions; it covered 30 percent of the group's GHG reductions. As in BP's case, the expertise gained through in-house trading made Shell a sought-after voice in the development of the EU ETS (Hoffman 2006). Yet the initial step from private in-house trading to the first major public scheme was taken in the United Kingdom. The UK ETG was the main player in devising the regime.

A Domestic Coalition: The UK ETG

In 1997, the New Labour Party rose to power in the United Kingdom, giving environmental issues new political prominence. Right after the Kyoto Protocol was agreed to, the UK government asked the Advisory Committee on Business and the Environment (ACBE) to provide advice on how business could contribute to the United Kingdom's response to climate change (Nye and Owens 2008). Launched in 1991, the ACBE had been an influential group of government-appointed senior businesspeople that shaped environmental policy and mobilized companies in support of environmental policy. In March 1998, the ACBE (1998) issued the report *Climate Change: A Strategic Issue for Business* in response to the government's request. The report argued clearly in favor of international emissions trading, while it also acknowledged a role for a carbon tax. The ACBE recommended that the government establish institutions for business-to-business trading as soon as possible. Further, the report suggested the implementation of a full-fledged domestic trading scheme to gain experience with carbon trading.

In the same month as the ACBE report was published, the government commissioned a task force led by Lord Marshall, then chair of British Airways, to consider whether emissions trading or taxation would be the more suitable economic instrument to regulate the use of energy. When the report "Economic Instruments and the Business of Energy" (Marshall

1998) appeared in November 1998, it essentially argued for a policy mix, including a tax and some degree of trading. Compared to the ACBE report, the Marshall report was considerably less supportive of emissions trading. The consultations in preparation of the Marshall report had shown that the energy industries and financial services sector were strongly in favor of carbon trading, while other sectors were hesitant. The Marshall report was particularly skeptical about how small and medium-size enterprises could participate in a trading scheme. In the end, the report did not back the ACBE's recommendation to develop a fully operational domestic trading scheme but instead suggested a "dry run" pilot scheme. The major recommendation for immediate implementation was a downstream tax on energy use.[4]

The Marshall report had significant impact on the United Kingdom's climate policy. In the budget of March 1999, the government announced the Climate Change Levy, a tax on downstream energy use that was to be implemented in 2001. The government also negotiated a number of Climate Change Agreements with individual industries. Nonetheless, the emissions trading proposal continued to have a sizable constituency among British industry and government. BP had already implemented its in-house trading scheme, and ACBE members were still favoring a trading scheme. In fact, "the concept of a UK ETS was already on the table prior to the Marshall Report . . . and [a number of companies] . . . were committed to pursuing emission trading as a long-term regulatory strategy for the energy industry" (Nye and Owens 2008, 11). To strengthen their political position, these actors set up the UK ETG.

On June 30, 1999, the UK ETG, a group of thirty organizations convening under the auspices of the Confederation of British Industry and the ACBE, was set up at BP's headquarters in London (interview 7). The founding members included, among others, BP Amoco, British Gas, National Power, and a number of industry associations, such as the Association of Electricity Producers, the Confederation of British Industry, and the ACBE (Rees and Evers 2000). The group tried to develop, in close collaboration with government officials, recommendations for a domestic trading scheme. The first draft proposal was announced in October 1999, followed by the second draft proposal in March 2000. "Unofficially, it [the UK ETG] was a politically well heeled advocacy coalition . . . with a core of elite business representatives determined to put emission trading back on the policy agenda after Marshall's less than favourable report" (Nye and Owens 2008, 5). Though a Europewide trading scheme was on the mind of the UK ETG members, they targeted the domestic level because they had "more political muscle" there, as one business representative stated (interview 52).

The UK ETG had started as a series of meetings between staff members from large emitters and government officials in the first half of 1999. On the business side, as one company representative put it, "BP did the phone arounds" to convince UK companies to join the group (interview 47). The majority of the early business members all came from the energy industries, including oil and gas producers as well as electric utilities.[5] On the government side, Henry Derwent from the Department of Environment assumed a leadership role. From the beginning, the UK ETG was a collaborative project between business and government, although BP assumed a particularly outstanding role. Chase became the UK ETG's steering committee chair. His executive assistant was named head of the UK ETG's secretariat, which—for the first few months—was located in BP's offices.[6] Another senior figure from BP was James Nicholson, who had already played a key role in the run-up to the Kyoto Protocol. In January 2000, Margaret Mogford was seconded by British Gas to become head of the UK ETG's secretariat, which by then had moved to the offices of Blue Circle, a cement company (interview 47). Another critical person was Bill Kyte from Powergen, an electric utility. Together these actors constituted the heart of "quite a powerful network," which might eventually also influence European climate politics (interview 7).

As Michael Nye and Susan Owens (2008) suggest, the UK ETG's advocacy for a domestic trading scheme was driven by both economic concerns about tax-based climate legislation and symbolic concerns about the role of business in environmental policymaking. In part, the UK ETG formed in opposition to a taxation agenda in general and the Climate Change Levy in particular (Brown and Taylor 1999). The majority of the UK ETG members did not qualify for a Climate Change Agreement, which would have granted them an 80 percent discount on the levy.

Given this circumstance, upstream energy providers were looking for options to reduce their compliance costs. Beyond the immediate costs of the Climate Change Levy, member companies wanted to set the agenda in favor of emissions trading, because they feared considerable cost from future climate legislation. A representative of the ETG secretariat phrased the rationale the following way:

> So, here are companies with big . . . exposures [to the levy], UK based, recognizing that these issues had to be taken seriously—that carbon control is a big issue and it will become a bigger issue in the future. And therefore that there was a

political and an economic game to be played. . . . It needed to be dealt with, it was a long-term issue, and what you needed to do is to begin to establish those institutions which would enable this issue to be addressed most effectively. (quoted in Nye and Owens 2008)

Next in significance to the anticipated cost of environmental regulation, the formation of a protrading group was spurred by industry's interest in reinstating its political influence vis-à-vis the government through political posturing. From early on in the process, a few energy companies had invested considerably in emissions trading as a regulatory approach. When the government announced the levy, industry felt that its efforts had not been acknowledged and its voice had not been heard. The energy industry's bid for a UK emissions trading scheme thus was motivated by the economic costs of immediate and, more important, future climate legislation as well as concerns about the political power of the industry.[7]

The UK government had its own motivation for supporting emissions trading. First, it was interested in proving that emissions trading did in fact work in order to set the European agenda in favor of emissions trading. Second, it intended to give London a jump-start on Frankfurt and Chicago in becoming the financial hub of the emerging carbon market (interview 47). The City of London subsequently adopted this strategy (Knox-Hayes 2009). A 1999 study by the Centre for the Study of Financial Innovation, which had been commissioned by the Corporation of London, found that London was in a "very strong position to provide markets for permit-trading at both a local and a global level" (Houlder 1999c, 9).[8]

As for environmental groups, they were not among the early advocates of trading in the United Kingdom. Members of the UK ETG had discussions about allowing NGOs as equal members. Yet UK environmentalists were—like most of their European colleagues—skeptical of emissions trading. Therefore, the UK ETG set up an NGO liaison group to familiarize environmental groups with the approach.[9] The World Wildlife Fund (WWF), Friends of the Earth, the Royal Society for the Protection of Birds, and others all participated in the group. Their acceptance for emissions trading grew, even though they had issues with how the UK scheme was designed.

In the end, the big emitters and the UK government were the driving forces behind the UK pilot trading scheme. Michael Meacher, environment minister, said: "The government welcomes the initiative. . . . We share business' aim of having a UK emissions trading scheme operational as soon as possible" (quoted in Cooper 1999a). He continued: "The scheme establishes the UK as a world leader in the field of greenhouse gas emissions trading" (quoted in Houlder 2001, 3). Following the ETG recommendations, the UK Department for the Environment, Transport, and the Regions (2000) published the consultation document "A Greenhouse Gas Emissions Trading Scheme for the United Kingdom" in November 2000. This was received with great interest around Europe, giving support to emissions trading as a real-world policy instrument (Zapfel and Vainio 2002).

The result was a compromise that designed a downstream cap-and-trade scheme based on the Climate Change Levy. The need to move downstream resulted from the levy, yet most of the ETG members were upstream energy providers. Hence, the compromise eroded the incentives for participating in the trading scheme. The UK ETS also did not offer relief to companies from the energy tax. The only incentive for voluntarily taking on an emissions cap was the payments that the government made available (a total of thirty million pounds). In fact, only a few companies signed on to the scheme, and trading volumes remained low. "The longer-term regulatory interests of a small number of powerful companies dominated the policy agenda surrounding UK emissions trading, pushing through a relatively weak and complicated programme that had little to offer mainstream business" (Nye and Owens 2008, 11).

The symbolic value of the UK ETS, however, was considerable. The UK ETG's advocacy had succeeded in putting emissions trading on the agenda of climate politics in the United Kingdom and the EU. The organizational dimension of business power is clearly visible in the case of the UK ETG. Through the aggregation of interests across energy and manufacturing industries, the group was recognized as a legitimate and collective voice with which the government could engage. This is reflected in the fact that the UK government delegated authority to the UK ETG to develop the trading scheme's rules. While government representatives participated in the work of the UK ETG, it was primarily a business group. Again, it is striking how central BP was to the process of organizing the advocacy coalition, with Shell increasingly assuming a similar role. The core of the protrading coalition was surprisingly small, which shows that a few corporate leaders can hold considerable leverage in organizing broad business collective action. When the UK ETG was launched, BP's Chase said: "I hope that as many firms as possible will seize this opportunity to be in on the ground floor of the world's first comprehensive greenhouse gas emissions trading scheme" (quoted in Houlder 2001, 3). Many more firms did indeed follow the path of BP and the UK ETG.

The United Kingdom had acted as Europe's backdoor for the import of market-based climate policy, with business acting as the transmission belt. It is not surprising fact that emissions trading entered Europe via the United Kingdom. A general affinity toward market-based instruments made UK industry and government receptive to the U.S. regulatory approach. In the words of Charles C. Nicholson (2003), group senior adviser as well as chair of the WBCSD's Climate and Energy Group: "Right from the outset the concept was better understood, and more widely accepted, in the UK both at [the] official and unofficial level. . . . Market mechanisms are viewed with less suspicion here. And business was an early supporter." In addition, the UK electricity sector's experience with trading in a deregulated power market added to UK industry's willingness to engage with emissions trading. The UK pioneers formed the nucleus of a broader coalition for an EU-wide trading scheme.

Building a Coalition for the EU Emissions Trading Scheme, 1998–2004

The bottom-up activities of business and member states triggered a top-down political process at the European level, ultimately resulting in the approval of the Emissions Trading Directive. In the immediate aftermath of the Kyoto conference, emissions trading was only one among several policy options available for the EU to meet its Kyoto reduction target of 8 percent. A number of member states, environmental groups, and industry associations remained opposed to the instrument. The question was whether a coalition for emissions trading would emerge in the face of the existing opposition. Surprisingly, UK energy companies and the European Commission managed to organize broad-based support for the instrument across key stakeholders, which led to the EU's U-turn on emissions trading.

The Coalition Makers: Oil and Power Companies

After the Kyoto conference, the EU was facing the task of designing a climate policy that would allow the union to meet its Kyoto target. From 1998 until 2000, the politics was about setting the agenda for a European climate policy framework. In this process, corporate trading pioneers from the oil, gas, and electricity sectors in the United Kingdom formed a close-knit coalition. This coalition teamed up with the European Commission and a few market-minded member states, such as the UK government, in promoting carbon trading as a core pillar of European climate policy. Through pilot projects and advocacy activities, the campaign helped to popularize emissions trading across Europe. Corporate leaders provided a proregulatory business constituency to enable public actors to go forward with emissions trading, to much the same extent as public actors were powerful allies for business. As early as January 1998, the corporate protrading lobby expressed its interest in a EU-wide trading scheme to European Commission officials (interview 8). Other actors, such as member states, environmental groups, and other industries were co-opted into the coalition over time.

The UK ETG and BP's campaign for emissions trading extended to European politics. While setting up its in-house trading scheme and developing the UK ETS, BP reached out to European business, commissioners, and reluctant member states such as Germany (interviews 7 and 47). A number of company representatives played a pivotal role in advocating emissions trading in Europe, including Chase, Nicholson, and Wriglesworth. Because these managers occupied key positions in the WBCSD and BusinessEurope, they were able to shape the perception of emissions trading among industry representatives.[10] Moreover, BP's in-house trading scheme had enhanced its legitimacy considerably with policymakers, as it showed leadership and demonstrated that international emissions trading worked (Victor and House 2006). Even before the publication of the green paper, BP expressed its support for a European trading scheme to influential officials in the European Commission (interview 48). BP representatives frequently mentioned that their leadership earned them a place at the table. Shell followed BP closely in establishing an in-house trading scheme and doing political outreach. As much as they competed with each other in becoming the premier climate leader, together the two companies were the corporate ambassadors for emissions trading in European climate politics.

Alongside the oil majors BP and Shell, a small group of power companies was "blazing the trail" for emissions trading, as one industry insider put it (interview 2; Houlder 1999a). Within Eurelectric, the association of European power generators, the UK members and Electricité de France were the real drivers of the regulatory approach—often against the opposition of German power companies (interview 2).[11] The Subgroup on Flexible Mechanisms, for instance, the internal group dealing with emissions trading, was chaired by Bill Kyte of Powergen, who also was a prominent figure in the UK ETG.

It is important to note that in the early stage, carbon trading was only championed by a small band of entrepreneurs from within the European electricity industry. Other electric utilities only accepted the instrument over time. Eurelectric ran the Greenhouse Gas and Energy Trading Simulations (GETS), a series of computer-based trading simulations of a trading scheme, starting as early as May 1999.[12] The idea was to see if emissions trading would work and how it would affect electricity companies. Knowing how it works was considered essential for advocacy to avoid the setting of rules, which would harm the power sector. Eurelectric sought advice from the International Energy Agency and BP.

Already after the first of four trading simulations, power companies were in favor of the instrument and concluded in their first report: "Free and open trading can help to meet emission objectives by lowering compliance costs and by giving a strong signal, via the price of CO_2 permits, on the economic implications of an emission objective" (UNIPEDE/Eurelectric 1999, 1). The simulation had helped to facilitate internal consensus on emissions trading, not least because power companies understood that they could pass on the trading costs to their customers, thus reaping extra profits (interviews 2 and 48). Henceforth, Eurelectric actively advocated the policy approach in the European political arena.

While GETS 1 included only electric utilities, GETS 2 covered all those sectors that would eventually be included in the EU ETS. Eurelectric felt it was important to make a case for an economy-wide scheme to ensure that the actual trading scheme sufficiently provided cost-effective abatement options, according to an industry insider (interview 2). The participation of energy-intensive manufacturing industries in the simulation actually increased the acceptance of the instrument among these actors. The GETS 2 report one year later concluded:

> The creation of a knowledge community between electricity utilities and industrial companies from 16 countries, with input from the European Commission, allowed considerable progress to be made. Gets2 could serve as a basis for the implementation of a voluntary market, compatible with voluntary sector-based reduction agreements, which would allow all participants to get ready, as from 2002, for the deadlines of 2005 (implementation of the European market) then 2008 (implementation of the world-wide market). (Eurelectric 2000, 60)

Thus, the simulation produced expertise about emissions trading and made industry familiar with the instrument. Next to the in-house trading schemes and the UK ETG, the GETS series was key in constituting a "knowledge community" among business. To draw the broad picture: key actors from the UK ETG—mostly oil and power companies—actively advocated trading in their respective associations at the EU level—namely, BusinessEurope and Eurelectric—in the attempt to create a business coalition for trading. Since these actors knew each other from the UK ETG, there was a high degree of informal communication, though no formal coordination of political strategies occurred.

Unlike their U.S. counterparts, European NGOs were not among the early supporters of emissions trading. The majority of environmental groups were skeptical of the instrument due to ideological preoccupations that turned them against market-based environmental policy.[13] In the aftermath of the Kyoto conference, European environmental NGOs went through a transformation, slowly starting to accept and then support emissions trading. The WWF changed its position right after Kyoto, Greenpeace accepted emissions trading but never became a strong advocate, while some branches of Friends of the Earth remained opposed to the approach. U.S. NGOs were aware of the difficult status of emissions trading in the European environmental community. Therefore, the NGOs tried to change the perception of emissions trading in Europe by organizing conferences. For instance, in 1999, Environmental Defense facilitated a Transatlantic Dialogue on Market Mechanisms in collaboration with the Pew Center on Global Climate Change (Petsonk, Dudek, and Goffman 1998).[14] At the agenda-setting stage, European NGOs were not yet part of the transnational protrading coalition. They would join the network by assuming the role of quality managers only at a later stage of European climate politics. For now, the Directorate General for the Environment of the European Commission assumed the role of environmental activist, emerging as the closest ally of the protrading coalition.

Public Ally: The European Commission

Business support for emissions trading coevolved with the emergence of the European Commission as the single most important public supporter of the instrument (Skjaerseth and Wettestad 2008). The relationship between the commission and the proregulatory forces in business was essentially reciprocal, as their agendas converged to some extent. In retrospect, a European Commission official said: "It would have been extremely difficult to be successful had it not been for allies in industry" (interview 48). The commission was not only responding to the business demand for an EU-wide trading scheme but also was very much an advocate itself. The policy entrepreneurship on carbon trading was essentially an instance of coleadership of the commission and a select group of firms. The leadership role of the commission is reflected by the fact that the Directorate General for the Environment under Wallstrom took the lead on climate policy.[15] The first step toward an EU-wide scheme occurred in June 1998, when the European Commission published the communication "Climate Change:

Towards an EU Post-Kyoto Strategy" (European Commission 1998), suggesting the implementation of a European emissions-trading scheme by 2005.[16] By then, the commission was already engaged in conversations with protrading advocates from the business community (interview 8). This step was followed by a communication in 1999, announcing a green paper for 2000 that would outline a European trading scheme.

The "Green Paper on Greenhouse Gas Emissions Trading within the European Union" (European Commission 2000a) was published in March 2000. The paper laid out the basic concept of emissions trading, and discussed sector coverage, allocation method, compliance and enforcement mechanisms, and the link to other policy instruments. The paper was met with broad-based support among the stakeholders. While industry in general was open to the instrument, a portion of the business community continued to prefer voluntary agreements at the time (European Commission 2001b). The paper marked a milestone in the EU's response to the Kyoto Protocol, putting the idea of a Europewide trading scheme firmly on the agenda of policymakers and managers. It was a bold move by the European Commission, given that the majority of member states were still skeptical of the policy instrument at the time (interview 8). Three key factors related to political opportunities at the European and international level account for the European Commission's decision to promote emissions trading within the union (Christiansen and Wettestad 2003). These differ from the U.S. administration's rationale for supporting emissions trading.

First of all, the commission's willingness to support emissions trading is a direct lesson from the "abortive attempt to introduce a carbon tax" in the early 1990s (interview 2). Since the competency for fiscal measures lies with the EU's member states, any tax proposal has to be unanimously adopted by the Council of Financial Ministers. Due to the strong opposition of industry and key member states, the proposal was never adopted, and the commission withdrew it in 2001. Jos Delbeke, head of the Climate Change Unit in the commission's Directorate General for the Environment, had been an influential architect of the carbon tax proposal. Having experienced the ferocious opposition to the proposal, he became a strong advocate of emissions trading, which was considered to be politically more opportune. In that respect, European industry determined the range of available policy options for a mandatory European climate policy: a tax was ruled out, while emissions trading had a business constituency. The commission acted within these constraints. Moreover, it had the competency to establish a trading scheme, which was not considered a fiscal matter. Wallstrom also quickly bought into emissions trading and strongly advocated the approach (interview 6). Given her business background, she was inclined to support market-based policies. Led by Delbeke, a number of commission officials who had been trained as economists were also in favor of market-based climate policy. They formed the Bureaucrats for Emissions Trading group (Skjaerseth and Wettestad 2008).

Second, the pioneering work of member states to develop domestic emissions trading schemes created pressure from below on the European Commission (Zapfel and Vainio 2002). In order to protect integration achievements with regard to the internal market and environmental policy, the commission had a vital interest in preventing the emergence of numerous incommensurable domestic schemes. The integration of the internal market for goods and services had taken fifty years. A fragmentation of the carbon market into several national markets was to be avoided by any means. Private actors in the United Kingdom strategically employed the pressure from below in a quasifederal system. The UK ETG aimed to make emissions trading popular in Europe In-house trading and national schemes were seen as pilot schemes for a European or global scheme. In that sense, UK business employed the lever of the EU's quasifederal system and the commission's interest in an integrated internal market to push emissions trading from the national to the European agenda.

In addition to these two EU-specific political opportunities, the development of international climate politics made a case for EU leadership on GHG trading (Hovi, Skodvin, and Andresen 2003). By 1998–1999, U.S. ratification of the Kyoto Protocol became increasingly unlikely, given the opposition to the treaty in the U.S. Congress. This led to doubts as to whether the Kyoto Protocol would ever enter into force. Yet it was clear to the European Commission that any future U.S. participation in a binding international climate treaty would be dependent on the inclusion of flexible mechanisms. Given the U.S. government's strong preference for the approach, it made sense to move forward with emissions trading, as it was likely to stay on the international agenda. When President Bush withdrew the United States from the Kyoto Protocol in 2001, the case for EU leadership on international climate policy became even more pressing if the Kyoto framework was to survive. Reflecting the EU's quest for global leadership, Wallstrom said in late 2002 when the EU Emissions Trading Directive was approved by the council: "The eyes of the world were on us." She also noted that "many American businesses have come to us and are eager for information about how our trading scheme will work" (quoted in Kirwin 2003, 10). Given its institutional power in the European policymaking process, the commission was a potent ally for business. The commission alone has the right of legislative initiative in the EU's three pillars—that is, those policy fields that have undergone European integration. This includes environmental policy. Hence, the agenda-setting power of the commission is extraordinary.

As for member states, most of them were undecided about what role emissions trading should play in European climate policy prior to the publication of the green paper. A few countries such as Germany and Finland actively opposed it, while the United Kingdom and Netherlands were in favor of the approach. The United Kingdom had declared making London the hub of the carbon market an official strategy. Unlike Germany, the United Kingdom had deindustrialized and had a prospering financial services industry, thereby well positioning it to play a crucial role in the carbon market.

Soon after the green paper's publication, the commission set up the European Climate Change Policy Working Group on Flexible Mechanisms. Meeting ten times between July 2000 and May 2001, the multistakeholder group was mandated to develop a regulatory framework for emissions trading, discussing the critical design components of a trading scheme (Watanabe and Robinson 2005). The thirty participants came from the commission, member states, industry associations, and environmental groups (European Commission 2000b). The following industry associations were represented: Eurelectric, the Federation of German Industries (BDI), the European Roundtable of Industrialists (or BusinessEurope), the European Chemical Industry Council (or the International Federation of Industrial Energy Consumers), and the UK ETG.[17] The group included major industry stakeholders from both camps—the proponents (UK ETG and Eurelectric) and the opponents of emissions trading (the European Chemical Industry Council [CEFIC] and BDI). The commission had carefully handpicked the members of the group to ensure representativeness (interview 48).

Environmental groups had not been among the early movers but were now gathering behind the emissions trading approach and assuming the role of quality managers that tried to ensure the trading system would deliver on environmental effectiveness. While they were no longer struggling with business over the policy approach, they had distinct preferences for the trading regime's design. After all, their interest was in environmental effectiveness, not primarily in cost-effectiveness. CAN Europe and the WWF, in particular, acquired expertise in emissions trading and made the case for the environmental effectiveness of the scheme (interview 41). The WWF said that the agreement on an EU emissions trading scheme would combine "environmental effectiveness with economic soundness," unless it was undermined by lax national allocation plans (quoted in Houlder 2003, 10).

Over time, participants from all three sectors—government, business, and the environmental community—developed an institutional interest in the group's work and started to advocate the regulatory approach among their respective constituencies (interview 8). It is interesting to note that the European Commission had a hand in facilitating collective action among key political constituencies by providing an institutionalized process. At a more general level, this directs attention to the role of state actors in organizing and aggregating societal interests. It also reflects the insight of the constitutive strand of neoliberal institutionalism, which says that institutions facilitate agreement among parties. In this case, those parties are not only governments but also firms and NGOs. The working group's final report in fact reflected broad consensus on moving forward with emissions trading in the EU:

> Working Group 1 recommends that emissions trading start as soon as practicable. Implementation of emissions trading within the EC should not wait for the progress made in defining the Kyoto mechanisms, and should be developed in the context of, and with a view to influencing the design of, an international scheme from 2008. A pre-Kyoto EC system should be viewed as a *learning-by-doing* process. (European Commission 2001a, 4)

At this stage of the process, the general principle of emissions trading was widely accepted among the central stakeholders as the main pillar of the future European climate policy. UK business had led the campaign for EU-wide trading and found a powerful ally in the Directorate General for the Environment, and vice versa. Together they persuaded key actors to step behind the proposal, also co-opting environmental groups into the coalition over time. A commission official said that persuasion was relatively easy because a good idea sells itself (interview 48). Emissions trading was such an idea—whose time had come. While agreement on emissions trading was broad based, it never was unanimous. Opposition came from two directions: German industry, and energy-intensive manufacturing industries.

Winds of Opposition: Germany and Energy-Intensive Industries, 1998–2004

Emissions trading did not rise to the top of the agenda of European climate policy without a struggle. German industry more generally and energy-intensive industries in particular had a hard time warming to the idea of a mandatory emissions trading regime (Skjaerseth and Wettestad 2008). While they achieved small victories in terms of concessions along the way, the proponents of trading won the upper hand regarding the choice of instrument.

The BDI and German labor unions were adamantly opposed to emissions trading, mainly for two reasons (cf. Christiansen and Wettestad 2003). First, German industry had achieved favorable negotiated agreements on CO_2 reductions, which it did not want to trade for policies that could entail penalties (German Industry 2001). Moreover, there was considerable reluctance within German industry to implement market mechanisms (interviews 2 and 7), which required that the government set a cap on emissions, an approach that was not in tune with neocorporatist business-government relations in the coordinated German market economy. Market-based environmental policy also met a certain ideological resistance in Germany toward "Anglo-Saxon" approaches (Svendsen 2002). In addition, German industry was skeptical of the bureaucracy related to emissions trading. Opposition to emissions trading, however, was not unanimous across German industry. The chemical industry and coal-based power generators were the most ardent opponents, strongly affecting the overall industry stance.[18]

Given German industry resistance, both BusinessEurope and Eurelectric had to deal with an internal "German problem" (Dombey and Houlder 2002; interview 2). It seems that in Eurelectric, trading advocates sidelined German power companies, making Eurelectric a frontrunner on trading. BusinessEurope instead appeared to settle on the lowest common denominator. It never became a herald for trading, as the commission received different messages from different people (interview 48). Notwithstanding German industry opposition, the German government declared its support of the proposed Emissions Trading Directive in December 2002, after other member states had made a number of concessions to German views.[19] This stance can be attributed to a domestic process and the intervention of transnational actors. The German ETG, a stakeholder group similar to the UK ETG or the commission's Working Group on Flexible Mechanisms, had been working to establish consensus on emissions trading.[20] In addition, on behalf of the UK government, BP lobbied members of the German parliament and government as early as mid-1999 to mobilize support for a European emissions trading scheme, thus "tipping the balance" to make Germany come around on market mechanisms (interviews 7, 51, and 52).

Next to German industry in general, energy-intensive manufacturing industries were a part of the industry opposition to emissions trading, with prominent actors being the European Lime Association, Eurofer (steel industry), and to some extent CEFIC. They voiced concerns about rising power prices due to emissions trading, which would compromise their international competitiveness (cf. Asselt and Biermann 2007). As the argument went, this could ultimately drive European manufacturing industry offshore. These industries thus tried to prevent the commission from moving ahead with unilateral action (interviews 7 and 11). When President Bush withdrew the United States from the protocol, energy-intensive industries in Europe jumped at the opportunity to express concerns about Europe taking unilateral action by implementing the EU ETS. Without the U.S. economy operating under carbon constraints, European industry could suffer from a decline in international competitiveness in the transatlantic marketplace. Bertil Heerink, an official with CEFIC, said: "The decision by the United States to drop out of the Kyoto Protocol is relatively new. There needs to be an impact assessment on competitiveness vis-à-vis the mandatory sectoral targets" (quoted in Kirwin 2001, 861).

In the end, energy-intensive industries could not prevail over the pro-regulatory forces. Several observers suggest that one reason for the failure of industry opposition to emissions trading was a strategic miscalculation (interviews 41 and 48). Antiregulatory business groups never expected emissions trading to actually be implemented, assuming it would be defeated in a same manner to the carbon tax. A previous lobbying success resulted in business overestimating its political clout. Therefore, energy-intensive industries did not organize collective action and thus lacked a single voice. As in the case of the United States and also the EU, the antiregulatory coalition showed less cohesiveness regarding the emissions trading proposal as compared to the carbon tax proposal. This suggests that the creation of coalitions indeed represents power through organization. The antiregulatory business coalition learned this the hard way. Only after the directive was agreed to did energy-intensive industries form the Alliance of Energy Intensive Industries to represent their interests more powerfully.

Despite industry opposition, member states reaffirmed their commitment to meet the EU's Kyoto target at the EU spring summit in Gothenburg in 2001. By that time, the proposal for a European trading scheme had sufficient political momentum to stay alive (Svendsen 2002). The political battle had moved on to the rules of trading: the policy as such was accepted, but its design was negotiated. Different factions of industry and green groups were fighting over whether the EU ETS would be a lax or stringent environmental regime (Debate on EU Emissions Trading

2003). The most controversial issues were the legal status of the regime, the allocation method, and the sector coverage. Environmental groups favored a mandatory scheme with auctioning as the allocation method, while industry lobbied for a voluntary scheme with grandfathering as the allocation method.

Until the adoption of the directive in 2003, the majority of business—including outstanding advocates such as BP—were calling for a voluntary scheme until 2008, when global trading was expected to begin (Commission on Collision Course 2001; interviews 2 and 8). Yet many companies and industry associations were aware of the fact that an effective trading scheme had to be mandatory. The policy statements of Europewide business associations did not reflect this awareness, because energy-intensive industrial sectors and German industry insisted on the scheme being voluntary. The commission instead had indicated from the outset that it was aiming for a mandatory scheme. In the end, the commission and environmental groups prevailed on this question.

Regarding sector coverage, the chemical and aluminum sectors were exempt from the scheme, which has been portrayed as a lobbying success of those industries (Markussen and Svendsen 2005). Insiders, however, contend that this was a conscious design choice on the commission's part rather than a response to industry pressure (interview 48). The resulting scheme included the electricity and heat, iron and steel, refining, chemicals, aluminum, minerals, and paper/pulp/printing sectors. The green paper had foreseen coverage of these sectors.

As for allocation, the commission had considered the auctioning of permits to industry. Yet the directive provides for 100 percent grandfathering—that is, the allocation of permits free of charge to industry, based on benchmarking or a historic baseline. This was the allocation method favored by industry (Markussen and Svendsen 2005). It also secured industry buying into the policy. In addition, industrial groups achieved the provision that the authority to allocate permits lay with member states. This provided companies and business associations with better opportunities for rent seeking in the allocation process than they would have had if the commission had settled the allocation of permits (Svendsen 2002). On both issues, sectoral coverage and allocation, the European Parliament was opting for the more stringent policy, yet was outweighed by the council and industry.

After the directive proposal had been published in October 2001, the final directive was adopted by the parliament and council in July 2003 (Rogers 2003), foreseeing the implementation of the EU ETS for January

2005.[21] At the press conference on July 2, 2003, when the parliament approved the directive, Jorge Moreira Da Silva, a member of the European Parliament, declared: "The carbon economy is born!" (quoted in Emissions Trading Breakthrough 2003). Industry and green groups welcomed the directive. In a letter to all the commissioners, Eurelectric (2003), for instance, congratulated them for their work on emissions trading. Environmental groups had grown into the role of quality managers. In a joint statement, CAN Europe, Friends of the Earth, the WWF, Greenpeace, and the Royal Society for the Protection of Birds said: "What is needed now are strict national caps on CO_2 emissions. Weak caps would punish those business that are already investing in clean technologies, while rewarding laggards" (Emissions Trading 2003, 56).

In an unusual demonstration of political will, the efforts of UK oil companies, the power sector, and the commission—the European triumvirate of market-based approaches—brought about the first cross-border trading scheme for GHGs. In this process of "multi-leadership" (Skjaerseth and Wettestad 2008, 191), business support for emissions trading created enabling conditions for government to enact mandatory climate policy. The relationship between the commission and carbon trading advocates from industry was one of partial interest alignment and mutual political support. The European Commission relied on industry backing to be able to go ahead with mandatory climate policy, given the abortive attempt of the carbon tax. Notwithstanding their lobbying successes regarding the rules of trading, antiregulatory business forces failed to organize collective action against binding emissions reduction mandates. Other actors, including member states, the environmental community, and other business actors, were co-opted into the emissions trading policy network over time. While business stood relatively united behind the policy instrument, business conflict existed over the rules of the game. Yet only in the implementation phase did this threaten the coalition's very existence.

The Future of Carbon Trading: The Coalition at Risk, 2005–2008

On January 1, 2005, the first multilateral carbon-trading scheme went operational, entering a trial trading period that ran until the end of 2007.[22] The implementation of the EU ETS shifted the political economy of carbon trading and thus the landscape of corporate interests. While business stood relatively united in the scheme's development phase, business conflict began to emerge once experience with real-world trading could be gained and the future design of the scheme was on the table. The existence of real gains and losses in the carbon market led both the winners and losers to raise their voices more vocally as to the scheme's future design. Furthermore, as negotiations on a post-2012 international climate regime were beginning, European business mobilized to ensure that emissions trading would be part of the global deal.

Extending the EU ETS: Lax versus Stringent Rules

Once the trial period of the EU ETS had begun, the political spotlight shifted to the design of the future regime.[23] So far the battle over a stringent versus a lax trading regime had only been fought between industry, on the one side, and environmental groups, on the other. With the advent of the financial services industry, however, a powerful force within the business community was taking sides with environmental groups. This increased the tensions within the coalition, in particular around the issue of auctioning. The first EU ETS trading period had been overallocated, meaning that too many permits had been given away and that it did not lead to a reduction in CO_2 emissions (Ellerman and Buchner 2007). This played into the hands of those groups questioning the environmental effectiveness of the scheme and demanding a more stringent design in the second trading period, including long-term reduction targets and a tighter allocation of emissions permits.

In May 2005, a few months into the operation of the EU ETS, the Corporate Leaders Group on Climate Change (CLGCC), a group of UK and international companies convened by the University of Cambridge Programme for Industry, called on the British government on the occasion of the G-8 summit in Gleneagles to adopt emissions reduction targets for 2025 to increase market confidence and reduce investment risks (CLGCC 2005). In June of the following year, the group petitioned the UK government to take on ambitious targets for phase 2 of the EU ETS (2008–2012) in order to provide incentives for investment in low-carbon technologies. James Smith, chair of Shell UK, said: "We want to work with the Government and other companies across Europe towards strengthening this critically important piece of policy. We need EU Governments to set clear targets for the ETS out to 2025 so that our businesses and others can have the confidence to make long-term investments in reducing emissions" (CLGCC 2006).

The call for tighter and long-term reduction targets was supported by the financial services sector. At the EU level, market intermediaries had so far been represented by IETA. As "a broad church," the association convened energy companies, industrial companies, and market intermediaries (interview 8). Once the market was up and running, and real business opportunities were at stake, though, the interests of the financial services sector increasingly diverged from those of the energy and manufacturing sectors, leading to tensions between those two groups in the IETA (interview 12). This led to the emergence of new business groups exclusively representing financial services providers. The UK-based group London Climate Change Services (now the Carbon Markets Association) spearheaded emissions-trading-related advocacy efforts in the financial services industry, which has been promoting the interests of market intermediaries since 2003–2004.

In October 2006, eighteen market intermediaries set up European Carbon Investors and Services (ECIS; now the International Carbon Investors and Services).[24] Shortly after its inception, the group publicly asked Manuel Barroso, the president of the European Commission, to significantly tighten the allocation of emissions rights for the second phase of the EU ETS. A spokesperson for the ECIS (2006) remarked,

> It is important that the EU does not over-allocate and it must set out caps that are consistent with the Kyoto Protocol otherwise no one will believe the EU's new targets to be agreed in 2007. Investors need credible and consistent political decision-making in order to deploy capital in the emerging carbon market. . . . [T]he Commission must stand firm, despite political pressure from its member states to back off. This will help create a realistic price for carbon which will be sufficient to reward investors in clean technologies and incentivise companies to cut their emissions. This is what everyone wants.

The UK government and the commission, in particular, heard this business call for tighter allocation. The commission rejected the national allocation plans of several member states including Germany (Trading Thin Air 2007).

As for long-term targets for emissions reductions, EU member states took decisive action at the spring summit in March 2007. Germany, which held both the council's and the G-8 summit's presidency, had ambitions to advance European and international climate policy. At the spring summit, the heads of state adopted the target to cut CO_2 emissions by 20 percent from the 1990 level by 2020. This target would be raised to 30 percent if other developed countries adopted similarly ambitious targets. NGOs and corporate leaders firmly supported the EU's move. NGOs had organized an unprecedented coalition with members from fields as diverse as environmental protection, development aid, and human rights to address European heads of state (EurActiv 2007). The EU CLGCC Group welcomed the targets of the EU member states, as long-term targets allow industry to defer a long-term carbon price, which increases certainty. Politicians were seeking the support from select industry groups to be able to legitimize their policy on not only environmental but also economic grounds.

The mainstream of business, however, received the announcement of the EU's target negatively. BusinessEurope said at a press conference that the carbon price would reach an "extremely high level . . . in case of far-reaching unilateral approaches" with "very damaging consequences in terms of macro and sectoral level" (quoted in ibid.). Hence, efforts to make the emissions trading scheme more stringent through tightening allocation and adopting long-term targets have led to a split within European industry, mainly between market intermediaries and a number of corporate leaders, on the one side, and energy-intensive industries, on the other.

This shift is significant, since it puts the coalition for emissions trading increasingly at risk, as insiders suggest. It might after all lead to a backlash by energy-intensive industries. Such tensions are typical for broad coalitions representing diverse interests: coalition partners agree on the basic principle, but not the details, as distributional battles are fought over the rules of the game. This demonstrates the fragility of business collective action in the face of competition among firms, economic sectors, and national economies. The protrading advocates only agree on the small common denominator of the idea of a low-cost regulatory option. Beyond that, there is a lot of disagreement over the design of carbon-trading schemes, because this has tangible distributional effects.

European Business Pushing for a Global Carbon Market

The EU ETS was set up for a trial period in an international trading scheme under the Kyoto Protocol. Thus, the scheme was supposed to be a stepping-stone to an international carbon market. Yet the Kyoto Protocol's first commitment period ends in 2012, leaving the future of international climate policy in general and international carbon trading in particular open. Given this uncertainty about the future of carbon trading on the international stage, a number of European business groups increased their advocacy efforts to ensure that a future global climate regime includes emissions trading. Though emissions trading had first been exported to Europe from the United States, it now became a European export product. Industry advocacy was aimed at "reinternationalizing" domestic regulation to avoid negative effects on its competitiveness, create investment certainty, and establish business opportunities in carbon and low-carbon technology markets. Reflective of the broader stance of European industry, Veroen van der Veer (2007), chief executive of Shell, wrote in the *Financial Times* in January 2007: "Policy makers the world over should make wider use of flexible market mechanisms such as the capping of carbon emissions and trading of credits that has occurred under the European emissions trading scheme since 2005. While Europe's CO_2 market is a good start, trading needs to become global in order to be truly effective." The main concern was that a policy gap in the regulatory framework for the international carbon market might appear post-2012, which would impair investments.

The EU-centered business coalition for global emissions trading emerged in the context of a number of policy processes that work toward a post-2012 climate regime. The UN negotiations on a new climate treaty were launched at COP 11, which also acted as the first Meeting of the Parties to the Protocol (MOP 1), in Montreal in December 2005. Running parallel to the UN track for a post-2012 regime, the G-8 summits increasingly became a policy forum to discuss the contours of a future international climate regime. In addition, a new political platform was created—the International Carbon Action Partnership. A number of industry initiatives targeted their lobbying activities at these processes and events. In January 2007, the Swedish electric utility Vattenfall launched the 3C Initiative, a global coalition of companies that advocates a post-2012 framework for global emissions trading. On the occasion of the G-8 summit in Heiligendamm, Germany, in June 2007, the 3C Initiative and a number of other business groups, including ECIS, the Business Council for Sustainable Energy, and the association of German power companies, pledged the G-8 members to agree "to maintain continuity in the legally binding frameworks underpinning the carbon market" (Business Community 2007).[25] The alliance also appealed to the G-8 leaders to agree at COP 13 in Bali to conclude negotiations on a successor agreement to the Kyoto Protocol by 2009.

At the Bali conference itself, this call was reiterated in the so-called Bali Communiqué, a statement produced by the UK and EU CLGCC, and supported by 150 international companies. The statement said:

> It is our view that a sufficiently ambitious, international and comprehensive legally-binding United Nations agreement to reduce greenhouse gas emissions will provide business with the certainty it needs to scale up global investment in low-carbon technologies. We believe that an enhanced and extended carbon market needs to be part of this framework as it offers the necessary flexibility, allows for a cost-effective transition and provides financial support to developing countries. (CLGCC 2007)

The Bali Communiqué stressed in particular that "the shift to a low-carbon economy will create significant business opportunities. New markets for low-carbon technologies and products, worth billions of dollars, will be created if the world acts on the scale required. In summary, we believe that tackling climate change is the pro-growth strategy" (ibid.).

The last sentence of the quote reflects that the actors firmly bought into the sustainable development agenda. While the majority of supporters were based in the EU, a number of companies came from the United States, China, and Australia. U.S. members included Coca-Cola, DuPont, Gap, GE, Johnson and Johnson, Nike, Pacific Gas and Electric (PG&E), Sun Microsystems, and United Technologies. The call for long-term clarity and a global framework for emissions trading was echoed by the WBCSD and the ICC (Harvey and Aglionby 2007). In the end, the Bali Action Plan, the outcome of COP 13, laid out a road map for negotiating a post-2012 regime, considering the use of market mechanisms. The future of international carbon trading and success of the reinternationalization efforts are not yet decided at the time of this writing. The point to be made, however, is that European companies are at the heart of the coalition for a global carbon market, which is unlike the first internationalization attempt in the run-up to Kyoto, when U.S. companies played the more prominent role.

The International Carbon Action Partnership, a new international organization promoting a global carbon market, was set up in October 2007 by countries that have either implemented or are aiming to implement mandatory cap-and-trade schemes (Ware 2007a). The partnership's goal is to develop a global cap-and-trade GHG emissions market through learning and linkage. Members include the European Commission, a number of EU member states, U.S. states that are members of the U.S. Regional Greenhouse Gas Initiative, U.S. states and Canadian provinces that participate in the Western Climate Initiative, New Zealand, and Norway.[26] Contrary to the UN track, which aims to establish a top-down framework for the carbon regime, the partnership pursues a bottom-up approach by linking existing and emerging carbon markets. In both cases the goal is a global carbon market. Referring to the globalization of emissions trading, Marc Stuart, the founder of EcoSecurities, concludes that "it would be a remarkable political turnaround to have got this far and see the whole thing disappear" (quoted in Wynn 2007).

Conclusion

The business coalition promoting emissions trading in the EU was essentially an alliance of big emitters. Large companies from the oil and gas industry along with the electricity sector in the United Kingdom pioneered the regulatory approach through in-house trading as well as a national trading scheme. In the entire process, BP acted as a key broker in building support for a Europewide trading scheme. In the United Kingdom as well as the European context, the protrading campaign was initially an antitaxation campaign. Firms, in this respect, were pursuing a risk management strategy. Interestingly, the antiregulatory forces were initially not organized as a coalition, which partly explains their relatively low level of influence.

The protrading coalition's influence instead manifested in a number of ways. First, the UK ETG firmly pushed back the carbon tax agenda in the United Kingdom and set up the UK ETS. UK oil and power companies were the only vocal constituents promoting this agenda. The activities in the United Kingdom had an impact on agenda setting at the EU level insofar as the UK ETG and in-house trading schemes created a "demonstration effect" for the feasibility of GHG emissions trading (cf. Green 2008). In addition, these activities generated expertise, which ensured access to the European Commission for trading pioneers. Without the support of a substantial portion of European business, the commission would most likely have failed to put mandatory climate policy in place. The history of European climate policymaking provides testimony for this, as the carbon tax proposal in the early 1990s was successfully defeated by business. Why did the corporate campaign for emissions trading succeed in the EU?

There was no immediate policy crisis in European climate politics that could be exploited by nonstate actors (see table 5.1). But there was considerable political will to implement the Kyoto provisions. By the late 1990s, the European public and policymakers had internalized a norm of climate action, which posed a threat to business rather than offering a political opportunity. Yet the protrading coalition could exploit two norm-related political opportunities with regard to their advocacy for carbon trading. First, the inclusion of flexible mechanisms in the Kyoto Protocol created international political momentum for the use of emissions trading in climate policy. Second, the liberal norms of the UK economy offered opportunities for advocacy for market-based climate policy. Emissions trading was a good fit with a liberal market economy as opposed to coordinated market economies, which are prevalent in continental Europe. Thus, the regulatory fit between emissions trading and a powerful national economy in Europe was essential to provide the regulatory instrument with an entry point to the European policy discourse.

The evidence that legitimacy was a significant source of influence lies in the fact that European Commission officials frequently pointed to the credibility of actors such as BP and Shell compared to much less constructive

Table 5.1
The EU ETS: The sources of influence of the Protrading coalition

Source of influence	Impact	Evidence
Political opportunities		
Policy crisis	Low	No immediate policy crisis
Norms	Moderate	Kyoto Protocol agreement created momentum for market mechanisms; liberal norms of UK economy offered opportunity for protrading advocacy
Coalition resources		
Financial resources	Moderate	Financial and organizational resources were important for in-house trading schemes, which created significant demonstration effects and enhanced the credibility of protrading advocates
Legitimacy	High	Legitimacy was the key resource in protrading advocacy; the comparative legitimacy advantage stemmed particularly from leadership and expertise
Coalition strategies		
Mobilizing state ally	High	European Commission was a powerful ally for the protrading coalition
Playing multilevel games	High	Coalition campaigned for emissions trading in key member states and at the European level; protrading advocates picked the most receptive member state for pilot trading scheme

voices from industry (interviews 8 and 48). BP, Shell, and electric power companies demonstrated goodwill toward the UK government and the European Commission in support of a business-friendly climate policy. This represented a departure from the adversarial business-government relations of climate politics in the early 1990s. Some companies supported emissions trading primarily for the sake of building good relations with government, which could be beneficial in other policy fields (interview 47). In addition to a cooperative attitude, some companies showed real leadership through the creation of in-house trading schemes and the adoption of voluntary emissions reduction commitments. This enhanced the credibility of BP and Shell in the eyes of the UK government officials and European Commission officials (interview 48).

While legitimacy was the political resource that ensured access to policymakers and influence in agenda setting, material resources played a role for business in becoming a legitimate voice. The in-house trading schemes of the two oil majors required the infrastructure of a multinational corporation to prove the feasibility of global emissions trading. Environmental groups, which lack such structures, could not have conducted such a policy experiment with a significant demonstration effect. This is a case in point that material forms of power can be essential in acquiring legitimacy.

As for the role of state allies, there exist striking parallels between the Kyoto and the EU cases. The European Commission played a part similar to that of the Clinton administration in setting the agenda for market-based climate policy. The commission is a powerful public actor, which holds an institutional preference for market-based instruments and strong political leverage vis-à-vis member states given its agenda-setting authority. This lends support to the notion that the rise of emissions trading in Europe is largely due to a convergence of some business interests and the preferences of a major political actor, the commission. It represents a case of the "multi-level leadership" (Skjaerseth and Wettestad 2008, 191) of business and a powerful supranational actor. A number of facts suggest that the emergence of such a state-business alliance relies on business organizing itself as a collective voice. The European Commission invited only associations, such as the UK ETG and BusinessEurope, to participate in the working group. Without organization as an advocacy coalition, UK trading pioneers would most likely have had less access to commission officials.

Finally, selling emissions trading in the EU was a multilevel, multitarget affair. The protrading advocates had to create support for emissions trading within key member states including the UK, Germany, France, and others. Given the distribution of political authority in European governance, a multipronged campaign was essential—critical battles were fought in the United Kingdom, at the European level, and in Germany. In addition, the campaign had to target a variety of actors beyond national governments, notably the environmental community, in order to affect normative change. A multipronged persuasion strategy was especially needed because of the ingrained opposition of European political actors to carbon trading. Both the distribution of political authority in European governance along with the diffuse nature of policy ideas and agendas called for a transnational campaign across multiple levels. Firms strategically shopped venues by first leveraging their clout at the national level, thus triggering action at the European level.

The creation of the EU ETS presented a significant milestone in the globalization of carbon trading, as it established the first multilateral trading scheme after the Kyoto Protocol had internationalized the policy idea of carbon trading. The formerly strongest opponent of emissions trading had become a champion of the policy instrument. This occurred during a period in which the U.S. government had largely withdrawn from international climate politics. Nevertheless, with the EU assuming a leadership position on climate change and setting the rules for a future carbon market, the United States, the homeland of emissions trading, was hard-pressed to catch up. The next chapter turns to the evolution of U.S. domestic climate politics in the post-Kyoto period, analyzing the role of business in reviving the carbon trading agenda in the United States.

6

The United States: Reimporting Carbon Trading

As the cradle of emissions trading, the United States was the main driving force behind its internationalization. After its withdrawal from the Kyoto Protocol, however, the United States refrained from the international political project of developing a global carbon market, settling instead for voluntary climate initiatives. Yet by 2009, a regional emissions trading scheme was operational, a large number of states were developing regional trading schemes, and the House of Representatives passed the first market-based climate bill. Carbon trading was back on the agenda. While chapter 4, among other things, explored shifts in U.S. foreign policy in the run-up to the Kyoto Protocol, this chapter focuses on shifts in domestic climate politics in the United States. The revival of carbon trading in the United States was very much a bottom-up process, which is why the following analysis cuts across political levels and political actors, even as it concentrates on the influence of business. Two questions are of interest: How did carbon trading rise back to the top of the climate agenda in the United States? What role did firms play in this process?

This chapter shows the evolution of business conflict and coalitions in the U.S. move from voluntary commitments to cap-and-trade bills in the post-Kyoto period. First, it will demonstrate that one of the last successes of the antiregulatory business coalition was the U.S. withdrawal from the Kyoto Protocol and the implementation of voluntary climate policies by the Bush administration. Then it will explore how from 2003 on, on-the-ground actors such as states and firms pioneered GHG emissions trading in the United States through the CCX and state initiatives in the Northeast and California. This analysis of corporate and state entrepreneurs is followed by a look at the emergence of deep-rooted conflict in the U.S. business community along with the creation and influence of a proregulatory NGO-business coalition, the U.S. Climate Action Partnership (USCAP).

When political parameters shifted in federal politics, the likelihood of mandatory federal climate legislation increased. In this situation, USCAP was launched to ensure that a cap-and-trade system would be the chosen policy instrument. The fourth section examines the effects of evolving business collective action on the legislative process in Congress. Cap-and-trade bills had been proposed as early as 2003, but momentum only picked up in 2007, putting it on the federal legislative agenda.

The Bush Administration and Voluntary Climate Policy, 2001–2003

In his election campaign, George W. Bush had pledged to enact mandatory reduction targets for the power sector. The president renounced his strategy shortly after taking office. Subsequently, he withdrew the United States from the Kyoto Protocol and began to promote voluntary industry activities. Though the transnational protrading coalition was gaining strength, the domestic antiregulatory coalition had won the battle over the U.S. ratification of the Kyoto Protocol.

The United States Withdraws from the Kyoto Protocol

In the early months of the Bush administration, officials reviewed U.S. climate change policy, while international negotiations were being delayed so that the United States could arrive at a negotiating position (Revkin 2001c). On March 13, 2001, President Bush sent a letter to four senators renouncing his campaign pledge to regulate the GHG emissions of electricity generators and restating his opposition to the Kyoto Protocol (Jehl and Revkin 2001). Two weeks later, Christine Whitman, the head of the EPA, announced that the United States would repudiate the Kyoto Protocol. While oil exporters such as Saudi Arabia welcomed the decision, most countries around the world reacted with dismay (Revkin 2001b).

To some the decision came as a complete U-turn by the United States on climate change, in particular because the protrading coalition was gaining ground. As Levy and Egan (2003, 825) observed: "It is somewhat ironic that the new Bush administration in the USA pulled out of the Kyoto Protocol in 2001, just as much of American industry appeared willing to accommodate mandatory international emissions controls." The Clinton administration had already failed to ratify the protocol, though. After all, the withdrawal from the protocol reflected the continuing power of the antiregulatory coalition including the GCC, right-libertarian think tanks, and Republican senators. Representatives of coal and oil companies as

well as electric utilities had lobbied the administration intensively right after it had come into office.

In a notorious two-page memo, Haley Barbour, a lobbyist for Southern Company, a coal-based power company, addressed several high-ranking officials.[1] Titled "Bush-Cheney Energy Policy & CO_2," the document said: "A moment of truth is arriving in the form of a decision whether this Administration's policy will be to regulate and/or tax CO_2 as a pollutant. The question is whether environmental policy still prevails over energy policy with Bush-Cheney, as it did with Clinton-Gore" (quoted in Newton 2002). Barbour criticized the notion of regulating CO_2 as "eco-extremism." As a former chair of the Republican National Committee and a Bush campaign strategist, Barbour had credentials that ensured he would be listened to by the administration. Moreover, Southern Company's donations to the Bush campaign were second only to Enron's (Revkin and Banerjee 2001). Both facts suggest that the memo played a key role in the administration's decision.

Next to the power industry, the oil industry in particular appears to have had extraordinary influence on the decision (Lisowski 2002; Goel 2004). While the industry has always been central in shaping U.S. domestic and foreign policy regardless of which party was in power, its influence within the Bush administration was especially strong. Bush himself had pursued a career in the oil and gas industry for eleven years. The fact that President Bush received US$1.9 million in campaign contributions from the oil industry underlines the close relationship. This sum is thirteen times the amount that Gore received in his 2000 campaign (Goel 2004, 483).[2] Having had its hand in Bush's decision, the antiregulatory business coalition welcomed the U.S. rejection of the Kyoto Protocol. GCC executive director Glenn Kelly said that the White House had received "a lot of communications. . . . Fortunately, the president responded quickly" (quoted in Jehl and Revkin 2001).

Members of the protrading coalition such as DuPont and Enron instead staunchly criticized Bush's withdrawal (Morgan 2001; Revkin 2001b). Bush's decision was seen as completely unhelpful (interview 33). An environmental expert of a large energy business said: "What businesses want is policy certainty. Bush has injected only turbulence" (quoted in Revkin and Banerjee 2001). While there was considerable business conflict over the decision given that the business community was divided into "two corporate camps" (ibid.), the protrading coalition did not actively advocate that the United States stick with the Kyoto Protocol.

A number of lessons regarding the influence and emergence of business collective action can be drawn from this. As for the question of influence, the failure of individual companies to prevent the U.S. withdrawal from the Kyoto Protocol suggests that concerted collective action is a necessary, though not the only, condition to gain weight in political decisions. Even though business was divided over U.S. foreign policy regarding the Kyoto Protocol, this conflict had little political effect because the pro-Kyoto group was too weak. Hence, business conflict without the collective organization of both parties tends to be politically meaningless. Another reason for the failure to prevent a U.S. withdrawal from Kyoto is the simple fact that the Bush administration was so clearly opposed to the treaty that no state ally for pro-Kyoto advocacy was in sight. Without some support from within the government, the protreaty forces in business could not change the policy preference of the White House, providing testimony to the crucial role state allies play in business advocacy.

What is more, a great number of the firms in the protrading coalition were not keen on mandatory climate policy in the first place, so the ratification of Kyoto was not necessarily in their interest. They would only promote emissions trading if the government was to push for emissions reduction mandates. This last aspect has interesting implications for thinking about the emergence of business collective action. Business might organize collectively not only if it detects political opportunities in the form of potential state allies but also when it faces a regulatory threat from governments. Under the Bush administration, such a threat did not exist, which adds to the explanation of why a protrading coalition did not emerge at this point in climate politics.

In sum, opposition from industry was a significant domestic reason for Bush's decision, although Michael Lisowski (2002) argues that other factors, such as the development of international negotiations, also played a part. The U.S. withdrawal from the Kyoto Protocol can be seen as the last big victory of the antiregulatory industry coalition, even as the protrading coalition was gaining momentum. Kyoto opponents were aware that it was not easy to kill the treaty. Again Kelly pointedly said: "The protocol is like the Titanic. After it sinks there are still going to be lifeboats that survive to be picked up by the next ship that comes along" (quoted in Revkin and Banerjee 2001). In the end, he would be right that the Kyoto approach had enough political stickiness to survive. Nevertheless, before key political parameters could change in favor of a mandatory emissions trading scheme, industry and the U.S. administration settled on the interim compromise of voluntary climate policy.

The Bush Administration and Voluntary Climate Policy

Following the U.S. withdrawal from the Kyoto Protocol, Bush turned to develop a domestic approach to climate policy around the voluntary activities of companies. Right after the announcement, the administration started to consult with an array of scientists, policy experts, and lobbyists on the future course of climate policy (Revkin 2001a). Representatives from both corporate camps—the protrading and antiregulatory coalitions—received access to policymakers. At this point, it remained an open question as to whether the result would be voluntary actions or a new binding policy.

The majority of opponents to Kyoto-style policies threw their political weight behind voluntary programs, as not taking any climate mitigation actions no longer seemed to be an option. The shifts in the political strategy of BP, Shell, and other corporate leaders had transformed the political terrain. The fact that they had acknowledged climate science and conceded the need for precautionary action changed notions of legitimate corporate conduct with regard to climate change (Levy and Egan 2003). This ultimately contributed to the GCC's decline. Instead of converting to the beliefs of European firms that mandatory limits to carbon emissions were inevitable, the firms defecting from the GCC entered into voluntary commitments and initiatives.[3] In addition, industry lobbied the Bush administration to launch voluntary initiatives that would defuse the public demand for climate action by promoting "no regrets" measures—that is, measures that are both environmentally and economically beneficial (Hopgood 2003). The Alliance for Climate Strategies in particular promoted such measures. Chaired by Kelly, the former GCC head, the alliance convened a number of former members of the GCC (interview 27).

In February 2002, Bush announced the U.S. Global Climate Change Policy, setting an 18 percent reduction target for GHG intensity to be met by 2012, which de facto implied an increase in absolute emissions (Krugman 2002). The Committee on Climate Change Science and Technology Integration had the oversight of the implementation of this goal, which was to be mainly achieved through promoting technology development and deployment as well as through voluntary commitments. The administration set up two multiagency programs, the U.S. Climate Change Science Program and the U.S. Climate Change Technology Program (Jones and McIntyre 2007). The program activities focused on advancing biofuels, clean fossil fuel and renewable energy technologies, and nuclear energy. US$4.6 billion of tax credits over a five-year period were meant to create incentives for technology take-up.

As for voluntary programs, the EPA's Climate Leaders program encouraged companies to set emissions reduction targets and annually report on their progress (U.S. Department of State 2007). More than a hundred companies from the manufacturing and energy industries have signed up for the program. In 2003, the Department of Energy launched Climate Vision, which engages trade associations of energy-intensive industries.[4] The groups each committed themselves to contribute to the president's overall emissions reduction target. Activities focused strongly on the development and commercialization of energy-efficient technologies. A number of other smaller voluntary programs, or public-private partnerships, complemented those two key programs.

While voluntary climate policy had become the dominant approach, the first bills suggesting a domestic cap-and-trade scheme were tabled as early as 2003. In August 2001, not long after Bush's withdrawal from Kyoto, Republican senator John McCain from Arizona and Democratic senator Joe Lieberman from Connecticut announced that they would jointly draft legislation on climate policy. Lieberman said that the administration's stance "abdicates the United States' position as a leader in environmental affairs and places U.S. industry at risk" (quoted in Lind and Tamas 2007, 103). Lieberman and McCain drew praise from Environmental Defense, whose Fred Krupp said that the senators' call for a cap-and-trade scheme "takes on the problem of global warming in a strong and sensible way" (quoted in Sullivan and Najor 2001). The response from business was mixed, with some companies opposing the senators' plan and others supporting it. For instance, Mark Carney of PG&E, a California utility, said that his company supported a cap-and-trade system (Cook 2001).

In 2003, the senators jointly introduced the Climate Stewardship Act, the first bill that set a limit on GHG emissions (U.S. Congress 2003a). The bill aimed at cutting the emissions of the electricity, transportation, industrial, and commercial sectors to 2000 levels by 2020 and 1990 levels by 2016. This goal would be implemented through a cap-and-trade scheme that allowed for the offsetting of reductions in developing countries and through carbon sinks.

The bill faced strong winds of opposition from the president, Congress, and business. President Bush insisted that his programs for voluntary action were sufficient and that the state of research would not justify further action. In the Senate, Republican James Inhofe of Oklahoma invited scientific skeptics of climate change to the Senate Environment and Public Works Committee, saying at the meeting: "With all of the hysteria, all of the fear, all of the phony science, could it be that man-made global warming is the greatest hoax ever perpetrated on the American people? It sure sounds like it" (quoted in U.S. Congress 2003b). A few corporate leaders backed the bill, but the major trade associations opposed it, supporting Bush's voluntary approach instead (Layzer 2007). These included, among others, the American Petroleum Institute, the Alliance of Automobile Manufacturers, the Edison Electric Institute, the American Iron and Steel Institute, the American Chemistry Council, and the Aluminum Association. The Edison Electric Institute and a number of other associations sent a letter to senators, claiming that "the bill's mandatory cap-and-trade and mandatory reporting approaches are the wrong ways to address global climate change" (cited in ibid., 115). In the end, the opposition prevailed, and the bill was defeated by a forty-three to fifty-five vote.

Voluntary measures had emerged as the domestic response to climate change. Yet more so in private than in public, many industry actors recognized that this could only be the approach in the short run. As E. Linn Draper Jr., chair and chief executive of American Electric Power, noted: "Eventually, you're going to have to have a hard cap of some kind" (quoted in Bradsher and Revkin 2001). To prepare for this eventuality, a small group of firms started to pioneer emissions trading. Everyone was clear about the fact that even though emissions trading could be done on a voluntary basis, effective schemes were intrinsically linked to a mandatory target-and-timetables approach to climate policy.

Pioneering Trading from the Bottom-up: Firms and States, 2003–2006

The regulatory lacuna at the federal level created space for actors on the ground to experiment with climate regulation in a bottom-up approach (Selin and VanDeveer 2009). Business and state initiatives especially became the laboratories for GHG emissions trading in the United States. While business and public actors engaged in these activities for different reasons, they both acted as transmission belts for the reimportation of emissions trading to the United States. Their activities can be understood as political entrepreneurship that helped to keep carbon trading on the agenda, although the time was not yet ripe for a broad protrading coalition to form.

Corporate Initiative: The CCX

The protrading coalition of business-minded NGOs and corporate climate leaders continued its efforts to keep emissions trading on the agenda of U.S. climate politics. The organizations involved developed internal GHG management systems, advocated market mechanisms, and demonstrated the feasibility of market-based climate policy through a private trading scheme. One crucial initiative emerged in the wake of Kyoto that tried to set the agenda for a national emissions trading scheme: the CCX.

As in Europe, in the United States too, the first real-world GHG trading scheme was developed by the private sector. The CCX, a voluntary trading scheme, was established between 2000 and 2002, with trading operations beginning in 2003. Centre Financial Products Ltd., a Chicago-based firm that had already played a key role in developing SO_2 trading in the United States, initiated the scheme.[5] The CCX was the brainchild of Richard Sandor, the chair and chief executive of Centre Financial Products, who had already promoted international GHG emissions trading at UNCED in 1991. The idea of a pilot trading scheme was put on the agenda even prior to Kyoto, when Sandor presented testimony before the Energy and Natural Resources Committee of the U.S. Senate in September 1997 and at the White House Conference on Climate Change in October 1997. He laid out what it took to develop a market for emissions credits, contending that "emissions trading is an environmental and economic winner. Trading must be near the top of the full menu of policies that will be needed to prevent the costly threat of climate change" (Sandor 1997, 4).

With a grant from the Joyce Foundation, Sandor and his colleagues started the development of the CCX in 2000. Forty-six entities participated in the design phase, including energy and manufacturing companies, offset providers, financial services providers, forestry companies, and two municipalities (Sandor, Walsh, and Marques 2002). The main advocates of emissions trading, BP and DuPont, were among the members, but also electric power companies such as American Electric Power and Cinergy, a large coal-based electric utility from the Midwest.[6] While initially only thirteen organizations engaged in trading, CCX membership grew to about three hundred in 2008.[7] Being a cap-and-trade scheme, the CCX required its members to make a legally binding emissions reduction commitment of 1 percent in 2003, 2 percent in 2004, 3 percent in 2005, and 4 percent in 2006, from a baseline determined by its average annual emissions levels between 1998 and 2001. Companies had various motivations to join the exchange. These included setting the agenda in the United States for emissions trading as well as acquiring knowledge about emissions trading, to be prepared in case of regulation was tabled, and selling credits.

The environmental community never endorsed the CCX; the exchange remained essentially a business project. In 2006, a coalition of nineteen environmental organizations under the leadership of Environmental Defense and the NRDC asked states and cities not to join the CCX. They argued that the lack of participation of environmental groups and other stakeholders in setting up the exchange made the scheme "an inappropriate model for cities and states that are essentially endorsing" it by participating. They also said that they "do not oppose CCX's efforts to recruit companies to their program as a learning experiment" (cited in Scott 2006), although they questioned whether the CCX would deliver emissions cuts, given its low emissions reduction target and loopholes in the rules. This controversy between business and environmental organizations reflects the continuous tensions between the two groups about the stringency of trading schemes, which is a persistent scheme in the politics of emissions trading.

Despite opposition from the environmental constituency, the CCX had clear public policy ramifications. When Congress started to engage with market-based climate bill proposals, the CCX frequently served as a role model and an example of corporate leadership, which will be discussed later. Yet it took critical parameters in federal politics to shift for the CCX to have some impact on domestic politics. In its early days, it instead represented an experimentation forum for political entrepreneurs on carbon trading.

The activities of the corporate sector on emissions trading extended beyond pilot trading to advocacy. In October 2000, Environmental Defense joined forces with a group of seven energy and manufacturing companies—Alcan, BP, DuPont, Ontario Power Generation, Pechiney, Shell International, and Suncor Energy—to form the Partnership for Climate Action. As a press release from Environmental Defense (2000) noted, "The primary purpose of the Partnership is to champion market-based mechanisms as a means of achieving early and credible action on reducing greenhouse gas emissions that is efficient and cost-effective." All member companies had adopted emissions reduction targets.

While this partnership was the first attempt to broker an NGO-business advocacy coalition for market-based U.S. climate policy, it was not a success story. Since the Republicans controlled Congress, there were no political opportunities to move U.S. climate policy forward. The international momentum for climate action also was not yet as strong as in later years when the Kyoto Protocol was ratified and signatories started implementing market-based climate policies. This is a case in point for the pivotal role that state allies and state action play in creating political opportunities for nonstate actor influence. At a more general level, the failure of the Partnership for Climate Action stresses the role that environmental factors have on where and when particular interest groups hold the power to affect political change. After all, it is about being in the right place at the right time. Given these limitations, the partnership focused more on sharing best practices among member companies and communicating these to the public (Partnership for Climate Action 2002; Interview 24). It was also used, according to one industry representative, as a showcase to demonstrate to the international community that the United States was not a hopeless case (interview 31).

Both the Partnership for Climate Action and the CCX made a case for emissions trading. The advocacy effort of the Partnership for Climate Action failed because there were no public allies that supported mandatory climate policies at the time. Industry and government had settled on voluntary climate policy. The CCX instead had a demonstration effect that played a part when mandatory climate policy was on the agenda. Yet the main role of these initiatives was that they kept a small community of corporate carbon-trading entrepreneurs alive by maintaining a network, from which a protrading coalition would later emerge. In parallel to private initiatives, individual U.S. states were fertile ground for the revival of the emissions trading agenda.

State Initiatives: The Northeast and California

U.S. state governments had been developing and enacting climate policy since the mid-1990s, but started adopting second-generation climate policies in the 2000s (Rabe 2004, 2006). Emissions trading has been playing an increasingly important role in such second-generation policies, while it is embedded in a policy mix including other policy instruments, such as fuel standards, energy-efficiency measures, and renewable portfolio standards (Pew Center 2006). In addition to these measures, states have initiated legal proceedings against the federal government because it refused to create national mandates for GHG emissions reductions. A few U.S. states thus took on the role of protrading advocates in their own right. Moreover, they became the first political venues for the successful advocacy of the corporate protrading alliance. While much of state-level action occurred parallel to changes in U.S. federal climate politics, it is widely understood that state action triggered federal political action. This is why I discuss state actions first before I shift the focus of the analysis to the federal level.

U.S. states emit GHGs in the order of the emissions of European countries and larger developing countries (Selin and VanDeveer 2007). Yet the political importance of state action does not necessarily lie in the effective reduction of GHG emissions; it is in the leverage effect it has on

U.S. federal climate politics (interview 49). State action creates significant political and institutional precedents that lead to the diffusion of environmental policies among states, which creates momentum for federal climate regulation. Put simply, state action can trigger a domino effect. With regard to emissions trading, two state initiatives have been setting the pace: the RGGI in the Northeast, and California's Global Warming Act. While many actors were involved in the development of these schemes, business saw state action as a window of opportunity to shape mandatory climate policy in the United States in favor of the use of market mechanisms.

In April 2003, New York governor George Pataki invited ten other governors from the Northeast to explore the possibilities of developing a regional cap-and-trade program for power plants. Massachusetts and New Hampshire had already capped the emissions of their own coal-power plants, while other states were considering similar steps. In addition, the six New England states had already committed in 2001 to reduce their GHGs to 1990 levels by 2010 and then to 10 percent below the 1990 levels by 2020. Given the relatively strong regional integration of the electricity market in the Northeast, policymakers considered a regional trading scheme more efficient than single state-based policies (Rabe 2006). After extensive debate and consultations with stakeholders, Delaware, New Hampshire, New Jersey, New York, and Vermont agreed in December 2005 to set up the RGGI (2005). Maryland, Massachusetts, and Rhode Island became members in 2007. The latter two had abandoned the RGGI in 2006 due to industry pressure. The Associated Industries of Massachusetts had opposed the RGGI from the beginning, arguing that it would be too costly to adopt carbon limits on the state level in the absence of federal action (Jones and Levy 2007).

The RGGI's creation is largely due to pressure from environmental groups and leadership by state governors. The NRDC pressured Governor Pataki to implement a cap-and-trade system for CO_2 after New York had already implemented such a scheme for SO_2 (interview 49).[8] Keen to maintain his green credentials, Pataki set up the Greenhouse Gas Task Force, which recommended a regional cap-and-trade scheme. Business was not among the early advocates of the scheme but rather responded to the governor's initiative. Many of the corporate leaders on the national and international stage were in favor of a national trading scheme. Moreover, the scheme only affected electric utilities, which is why only a limited number of firms engaged with the political process. The business response was divided along the lines of fuel mix: nuclear energy and natural gas providers were in favor of a regional emissions reduction mandates, while coal-based utilities opposed it. The Nuclear Energy Institute primarily represented the former group's interests, while the New York State Coalition of Energy and Business Groups and the EEI were the voices of the latter group.[9]

All the opposition's efforts to derail the RGGI process failed. The political battle was mainly about the allocation method and other design elements. None of the groups involved ever questioned the use of market mechanisms. The antiregulatory groups opposed carbon caps, not market-based policies as such. BP was quietly involved in the RGGI process through its membership in the Greenhouse Gas Coalition. Though the company does not have any energy-intensive facilities in the Northeast, it wanted to make sure that the RGGI would set a precedent for a federal trading scheme that was favorable to the company (interview 52).

The fundamental design of the trading scheme was outlined in a "model rule," which was finalized in August 2006. Subsequently, the model rule had to be implemented by all participating states. The RGGI will eventually establish a regional emissions inventory, registry, and trading mechanisms for CO_2 emissions from the power sector. In January 2009, the RGGI became the first operational public carbon-trading scheme in the United States.

The northeastern states were soon followed by California and other western states. In June 2005, California governor Arnold Schwarzenegger signed an executive order that set several emissions reduction targets, including the goal to reduce GHG emissions to 2000 levels by 2010, 1990 levels by 2020, and 80 percent below the current levels by 2050 (Rabe 2006).[10] In response to Schwarzenegger's executive order, Democratic assembly member Fran Pavley introduced a bill proposal that resulted in the California Global Warming Solutions Act (AB 32) in August 2006 (California Assembly 2006).[11] Unlike the RGGI, AB 32 provides for an economy-wide and legally binding emissions reduction target. The bill foresees implementation measures to be in place by 2012. California's leadership led other states in adopting emissions reduction targets. Shortly after Schwarzenegger's executive order, New Mexico governor Bill Richardson announced comparable reduction targets. While the New Mexico act does not mandate a cap-and-trade program, it says that implementation may occur through a cap-and-trade scheme. Pavley had remained largely agnostic about emissions trading. The actual inclusion of the instrument in AB 32 is the result of a political battle between a set of players unique to California.

Market mechanisms were supported by, first and foremost, the governor, a few business groups and companies, and a number of environmental organizations, such as Environmental Defense and the NRDC. The latter two officially sponsored the bill from November 2005 on. Only a small band of corporate leaders such as the utility PG&E, Waste Management, Environmental Entrepreneurs, and eventually the Silicon Valley Leadership Group clearly supported AB 32 (Whetzel 2006).[12] Peter Darbee, chair, chief executive, and president of PG&E, put his weight behind state-level regulation in 2005, stating: "The incentives really aren't there for the creation of new technologies and investments to reduce carbon dioxide unless mandatory caps are put in place. Now, that creates an element of certainty" (quoted in Mouawad and Peters 2006). Both the Environmental Entrepreneurs and the Silicon Valley Leadership Group have been stressing the economic opportunities from climate policy for the clean-tech sector. The two business groups and the NRDC jointly published the report *A Golden Opportunity: California's Solutions for Global Warming* in June 2007. The bill also received considerable general support from faith-based groups.

Other business groups and companies such as the ICCP and BP got involved, but they mainly tried to ensure the inclusion of market mechanisms in the bill, even though they were generally not in favor of state-level action. Yet they were aware that California's climate policy framework would set a landmark precedent for any federal climate policy to come (interviews 31 and 50). Along with these companies, Kevin Fay, executive director of the ICCP (2006), pushed for market mechanisms, while expressing the ICCP's concern about the inefficiencies of state-level action.

Opposition to market mechanisms came from the environmental justice movement and legislators as well as antiregulatory business groups. Historically a strong player in environmental politics in California, the environmental justice movement was concerned that a cap-and-trade system might disadvantage poor urban communities by leading to a concentration of pollution in so-called hot spots. These had been observed in RECLAIM, a cap-and-trade scheme established by the Los Angeles area's air quality management authority in 1994 to reduce non-GHG pollutants. "We're skeptical of how efficient and just a cap-and-trade system can be because RECLAIM resulted in the concentration of pollution in low-income communities," said Philip Huang, a staff lawyer for Communities for a Better Environment, a statewide group that represents working-class and poor urban areas (quoted in Collier 2007). A number of Democratic legislators in the California Assembly were also skeptical of emissions trading, having had a long-standing preference for command-and-control regulation (Yi 2007). They aligned themselves with the environmental justice movement. While the movement was in favor of climate legislation but questioned the social sustainability of cap and trade, a number of major business associations rejected emissions reduction mandates as such.

The California Chamber of Commerce led the opposition to the bill, arguing that AB 32 would jeopardize the competitiveness of California's economy. "In effect, AB 32 imposes a new tax on business only in California, especially through the so-called *cap and trade* provisions. When businesses that have nothing to trade buy credits, they essentially are buying a permission slip to continue employing Californians and generating tax revenue for the state," said Allan Zaremberg, the chamber's president and CEO (quoted in California First 2006). Other groups opposing the legislation included, among others, the Western States Petroleum Association, the Manufacturers and Technology Association—which feared a rise in power prices—and the California Nevada Cement Production Council (Whetzel 2006; interview 50). It is important to note that the environmental justice movement's opposition was directed at market mechanisms, whereas antiregulatory business groups fought the general notion of regulating GHG emissions.

While the question of whether mandatory climate policy would be enacted was decided when the governor issued his executive order, the question about the role of cap and trade was resolved only when Governor Schwarzenegger insisted on market mechanisms in summer 2006. While the governor had always leaned toward a market-based approach, a climate change meeting in July 2006 is said to have been a turning point for him (interview 50). The Climate Group, an international NGO bringing business and government together, and BP organized a roundtable discussion between British prime minister Blair, Schwarzenenegger, and a select group of CEOs at BP's facility at the port of Long Beach, California (BP 2006).[13] This offered Schwarzenegger an international platform to demonstrate world leadership on climate change. Supposedly, Lord Browne's account of BP's experience with emissions trading convinced the governor (interviews 50 and 52). The two leaders had a one-hour private conversation before the official event. At the event, Blair and Schwarzenegger also signed a statement of intent to develop a transatlantic market-based framework to reduce GHGs (Wintour 2006). Steve Howard, CEO of the Climate Group, an international NGO working on climate policy, and

Browne of BP brokered the deal. The bill was passed a month later, including the option to implement it through a cap-and-trade scheme.

In a strategic move to strengthen the market-based approach, the governor established the Market Advisory Committee after signing the bill. The committee provided expertise on the design of cap-and-trade schemes to the California Air Resources Board. It consisted of national and international experts, including experts from Europe and the European Commission who have experience in designing cap-and-trade schemes, such as the EU ETS, the Acid Rain Program, the NOx Budget Program, and the RGGI. The advisory body published its first report, *Recommendations for Designing a Greenhouse Gas Cap-and-Trade System for California* (Market Advisory Committee 2007), in June 2007, making a strong case for a statewide emissions trading scheme, which would be linked to other mandatory systems such as the EU ETS and the RGGI in order to promote a global carbon market. Moreover, in another strategic action to promote the creation of a carbon market, Schwarzenegger set up the Western Climate Initiative (2007) with the governors of Arizona, New Mexico, Oregon, and Washington State in February 2007. Later in the year, the governor of Utah and the premiers of the Canadian provinces British Columbia and Manitoba joined the initiative. The objective is to reduce regional emissions 15 percent below 2005 levels by 2020 by means of a regional cap-and-trade scheme.

State leaders had indeed triggered a domino effect. States provided an entry point for public carbon-trading schemes to reach the United States. Alongside like-minded public actors, key members of the protrading business coalition had fought for the use of market mechanisms in state climate policies. They had also established a network of corporate lobbyists and state-level policymakers in favor of carbon trading, from which a federal coalition could emerge. But the activities of U.S. states also stirred fears among managers of a regulatory balkanization of the United States. With states thus raising the pressure for federal action on climate change, "the centre of gravity [of climate politics] shifted to Washington, DC" (interview 50). The temporary compromise solution of voluntary climate policy was being challenged—not least by business itself.

Business and NGOs Calling for Cap and Trade, 2005–2008

Business and state activities from below had increased the momentum for national climate regulation. When crucial political parameters changed in national politics, a window of opportunity was created for a mandatory federal climate policy. A number of companies and environmental groups jumped at the opening by establishing the U.S. Climate Action Partnership, a broad NGO-business coalition supporting a federal cap-and-trade scheme.

Political Shifts and the Role of Climate Leaders

Business and NGOs did not mobilize in response to a distinct event. Rather, the convergence of a number of political developments suggested that the temporary compromise of voluntary climate policy was increasingly at risk. Depending on the perspective, this was perceived as an opportunity or a threat (interviews 24, 40, and 44). Out of four key developments, three relate to domestic politics, while one refers to international politics.

First, as discussed earlier, California and states in the Northeast were taking action, which signaled an increasing political will among policymakers and the U.S. public to adopt emissions controls. State action put pressure on the administration because multiple regulatory frameworks increase compliance costs significantly for companies operating nationwide.

A second trend concerns the activities in Congress. Members of Congress were getting increasingly more active on the climate change issue, responding to pressure from the grass roots. A revised Climate Stewardship Act was introduced in spring 2005. In early June, shortly before the final Senate vote on the Climate Stewardship Act, four executives testified in front of the House Science Committee. In his written statement James Rogers (2005, 17), CEO of Cinergy, said:

> I believe that the country needs leadership in this area. I don't believe that I am being disloyal to the President whom I support, to Congress or to my shareholders when I say that the time is now to move positively toward reachable goals that will not only put us on track to operate in a greenhouse constrained environment, but on a track that will also make this country the technological leader it once was and can be again.

Though corporate leaders became vocal on climate policy, the Climate Stewardship Act lost the second Senate vote by thirty-eight to sixty on June 22. 2005. Yet on the same day, the chamber approved a resolution in the context of the debate about an amendment to the Energy Act (interview 40). The resolution said: "It is the sense of the Senate that Congress should enact a comprehensive and effective national program of mandatory, market-based limits and incentives on emissions of greenhouse gases" (U.S. Congress 2005, sec. 1612).

The momentum for federal climate action was clearly growing. In addition, the midterm elections in 2006 produced a Democratic majority in both houses of Congress. Democrats were more likely to enact mandatory climate policy than Republicans. And the balance of power was not only shifting within Congress but also between Congress and the White House. Generally speaking, the balance of power shifts more toward Congress the longer a president is in office (interview 25). Market-based climate bills were introduced mainly during President Bush's second term of office.

A third trend lay in the fact that the media and general public had awoken to the climate change issue due to, for instance, Hurricane Katrina and Gore's movie, *An Inconvenient Truth* (Hertsgaard 2006). Both events have been frequently cited as "shocks" that raised public awareness and mobilized the willingness to take serious action on global warming (interviews 21 and 32).

Finally, as a fourth trend, international progress in climate politics increased the pressure on the United States (Bang, Tjernshaugen, and Andresen 2005). After Russia had ratified the Kyoto Protocol, the treaty entered into force in February 2005. This left the United States increasingly isolated in not taking mandatory action on climate change. It also stimulated activities on the ground, as progress in international climate politics created a norm of climate action (Selin and VanDeveer 2007).

These domestic and international political developments, in combination, suggested an increased possibility for federal climate legislation in the United States, which brought companies back to the table. Unlike in any previous phase of U.S. climate politics, from 2004 onward climate regulation was looming over business as a realistic threat. This ultimately led to a deep split within U.S. industry over political strategy—a strategy that had been pioneered by a few corporate leaders. In more than one way, the struggles inside the U.S. business community in the run-up to Kyoto repeated themselves, though in a different political climate. The second round of the climate battle was on.

Against the backdrop of a shifting political context, a number of corporate leaders started to demand legislation, leading to a divide in the U.S. business community, which later led to the creation of a protrading NGO-business coalition. Next to the long-standing leaders on climate change such as DuPont and BP, electric utilities, technology companies, and financial services firms joined the call for federal climate legislation (Goodfellow 2005). In its annual report, the Carbon Disclosure Project noted that "a sea-change in corporate positioning on climate change is discernible. . . . Perceptions are changing most noticeably among U.S.-based companies, many of which have publicly asked for greater regulatory certainty on greenhouse gas emissions" (Innovest 2005). Two companies played especially prominent roles: Cinergy/Duke Energy and GE.[14] They represent the mobilization of large emitters, the electric utilities, and the potential winners from a price on carbon. In 2004, Cinergy took a public stance in favor of mandatory climate regulation and emissions trading, and in May 2005 GE joined in the call for action. Both companies would later be at the heart of USCAP.

In 2004, Cinergy (2004) published a report that made the case for a cap-and-trade scheme. This move was the result of two shareholder resolutions that were filed by the Committee on Mission Responsibility through Investment of the Presbyterian Church in 2003 and 2004, respectively. These two resolutions asked Cinergy to consider the potential implications of federal carbon constraints on its business. Rogers, Cinergy's CEO, had already become interested in the climate change issue after returning from the World Economic Forum in Davos, Switzerland, in 2000. In response to the shareholder pressure, he decided to take a leadership position on climate change, arguing that "if you're not on the table, you'll become the menu" (quoted in Helm 2005). Rogers had already supported emissions trading to solve the acid rain problem (Thompson 2008). He liked the elegance of the market-based approach and believed the technology would be found to respond to the problem. The fact that PSI Energy, the electric utility he was heading at the time, could cost-effectively comply with the regulation gave him the faith that a market-based approach works. Next to giving speeches at conferences, Rogers (2005) also testified before Congress in June 2005, demanding action on climate policy.

When Cinergy merged with Duke, the new company Duke Energy adopted Cinergy's policy position. Shortly after Cinergy had issued its report, American Electric Power stepped forward, advocating "comprehensive, cost-effective public policies that facilitate prudent, near-term emission controls" (Cogan 2006, 62). Rogers and Duke Energy continued to play an outstanding leadership role in corporate advocacy for climate regulation. Eileen Claussen of the Pew Center acknowledged Rogers's contribution: "It's fair to say that we wouldn't be where we are in Congress if it weren't for him. He helped put carbon legislation on the map" (quoted in Thompson 2008).

Next to the strategic shift of a few electricity companies, GE announced a high-profile investment and public relations initiative on environmental technologies and climate change in May 2005. Through the Ecomagination initiative, GE committed itself to an emissions reduction target as well as doubling its research budget for energy and environmental technologies to $1.5 billion by 2010. The initiative was preceded by an eighteen-month development phase, in which GE consulted with customers (Lean, Clean Electric Machine 2005). In a consultation session with executives of electric utilities in July 2004, Jeffrey Immelt, the CEO of GE, told his customers that limits on CO_2 emissions were to come. He noted that CEOs would be well advised to participate in making those rules (Kranhold 2007).

The initiative was embedded in a push for regulation to increase demand for environmental technologies. At the public launch of Ecomagination at George Washington University on May 9, 2005, Immelt (2005, 7) said: "We are living in a carbon-constrained world where the amount of CO_2 must be reduced. . . . We think that real targets, whether voluntary or regulatory, are helpful because they drive innovation. . . . We believe in the power of market mechanisms to address the needs of the environment." At this event, Jonathan Lash, WRI's president, welcomed GE's initiative. Moreover, the idea of an NGO-business coalition was first discussed at the Ecomagination's launch (interview 44).

Finally, major financial services firms joined in the call for federal action on climate change. The capital markets awakened to the climate change issue because they started to see the risks and benefits that climate policy would bring to their customers. In addition, when the EU ETS moved into its implementation stage, banks realized the potential benefits from carbon trading. In April 2005, the U.S. bank JPMorgan Chase stated that it would be initiating "a policy dialogue to advocate that the government adopt a market-based national policy on greenhouse gas emissions" (cited in Goodfellow 2005). The company was soon followed by the investment bank Goldman Sachs, which released a new environmental policy in November 2005, calling on the U.S. government to adopt a "strong policy framework that creates long-term value for GHG emissions reductions"(quoted in Cogan 2008, 20). These early corporate supporters were exceptions to the mainstream of business, which supported voluntary initiatives in line with the Bush administration's stance. While they represented a minority, their leadership ultimately led to a major split within the U.S. business community over climate policy when the proregulatory forces organized politically.

The pioneering role of a few companies defecting from the mainstream might suggest that business conflict was sufficient for the balance of power to shift in favor of cap and trade. The leadership of a few companies was crucial for business collective action to emerge, but it was insufficient to cause a change in government policy. Again, business conflict and business collective action are intrinsically coupled. The disagreement of firms over political strategies creates opportunities for the emergence of a new collective force. When this force organizes, business conflict becomes a politically relevant conflict. In this sense, business conflict is often the first step in the associational organization of business, which means that in the wake of business conflict, we often observe the emergence of new associations or alliances, which then might tip the balance toward a new business agenda. This was the case in the United States.

The Emergence of USCAP

On January 19, 2007, USCAP, a coalition of initially nine companies and four environmental NGOs, was launched.[15] Collectively, USCAP (2007c) members had total revenues of $2 trillion and employed 2.5 million workers, reflecting the economic power of the group. USCAP's launch significantly shifted the tectonics of business advocacy in U.S. climate politics. For the first time in the history of U.S. climate politics, an alliance of multinationals and environmental groups was demanding mandatory emissions controls through a cap-and-trade scheme.

The partnership was a cross-sectoral alliance, whose member base consisted of major companies from most economic sectors, including the oil, power, automobile, mining, and chemical sectors. Long-standing advocates for emissions trading such as BP and DuPont also participated in the coalition. Unlike in the preceding cases, USCAP gathered both big emitters and corporate winners of climate policy. Accordingly, USCAP had a twofold agenda. On the one hand, big emitters in the coalition pursued a proregulatory risk management strategy to contain compliance costs. They did so by advocating a cap-and-trade scheme and trying to influence the rules of the game (e.g., free allocation and price containment). On the other hand, winners including GE (energy-efficient technologies) and DuPont (biofuels and credits for early action) pursued a proregulatory market-making strategy once the likeliness of federal climate legislation was high.

As for NGOs, Environmental Defense, the WRI, the National Wildlife Federation, the NRDC, and the Pew Center on Global Climate Change were involved. By 2007, as USCAP's membership shows, the partnership's pragmatic and cooperative approach had become mainstream among U.S. green groups. This differed starkly from the situation in the mid-1990s, when the majority of environmental organizations pushed for targets, timetables, and command-and-control policies. Environmental Defense had essentially broken ranks with its peers—similar to BP in the oil industry—by promoting market-based climate policy and cooperating with business (Pulver 2004; Alcock 2008). As a representative of a key environmental group in USCAP pointedly expressed the new pragmatism: "The perfect is the enemy of the good" (interview 34).

In its "Call for Action," USCAP states that climate science is clear, which stands in direct contrast to the GCC's strategy a decade earlier. Furthermore, the coalition calls for a federal cap-and-trade scheme in the United States (USCAP 2007a). Krupp commented that "with this line-up of companies and environmental groups endorsing it, a carbon cap is clearly the consensus solution to climate change. . . . We chose a cap-and-trade approach because it guarantees the emissions cuts we need, while it unleashes cash and creativity from the private sector" (Environmental Defense 2007). USCAP members also published a letter in January 2007, asking congressional leaders and President Bush to take quick action on mandatory measures. The letter asserted that USCAP "is united in the belief that we can, and must, take prompt action to establish a coordinated, economy-wide, market-driven approach to climate protection" (cited in Scott 2007a).

USCAP emerged from conversations begun at the launch of Ecomagination in 2005 between WRI president Lash, Environmental Defense president Krupp, and GE chief executive Immelt (interviews 20 and 44). The organizations had preexisting relations; for instance, Lash had worked with GE on Ecomagination's launch (Barringer 2007). In January 2006, the coalition had its constitutive meeting, and in April of the same year the coalition partners started developing a policy position (interview 40). Over the course of the summer, USCAP expanded its membership to thirty-three organizations, including twenty-seven companies and six environmental groups. Key actors in the coalition, such as GE's Immelt, reached out to other firms to persuade them to join USCAP. But firms also applied to join the coalition, following a "herd mentality," as one participant said (interview 24). Companies pursued a number of goals by participating in USCAP: ensuring the economically most-efficient policy solution would be chosen (e.g., BP), receiving credits for early action (e.g., DuPont), increasing demand for a company's own technologies (e.g., GE), knowing what the playing field will be (e.g., Duke Energy), and getting certainty for a company's investments (e.g., Duke Energy) (interview 44).

The relatively high level of legitimacy of USCAP vis-à-vis policymakers was mainly due to two factors. First, USCAP tried to strike a careful balance between a large number of players to enhance the representativeness of the alliance, on the one hand, and keeping its membership at a manageable size, on the other (interview 44). To achieve this goal, USCAP convened large companies from a broad range of industrial sectors. This ensured representativeness by providing a cross-section of the economy instead of gathering as many companies as possible. Several of USCAP's members also were extremely successful economically, which increased the coalition's legitimacy, as these companies made a case in point that both environmental protection and business success could be achieved. GE, for example, is the world's largest corporation and is widely admired by executives for its long-term success.

Second, the cooperation between business and some of the largest, most well-regarded environmental organizations generated legitimacy. There are a couple of reasons why cooperation by two antagonistic groups increases the legitimacy of a political demand compared to when business goes it alone. Environmental groups endorse the environmental effectiveness of emissions trading, while business support for the instrument suggests that it is cost-effective and offers new business opportunities. Thus, the coalition itself represents the win-win logic of ecological modernization. Since business and green groups are the two primary constituencies for policymakers in environmental politics, joint action by these groups provides a strong case for government action along the lines of their proposal.

In developing a position, USCAP members only focused on the economics of mandatory climate policy and different design choices. It is interesting to note that USCAP did not discuss the state of climate science, which represents a clear break with any previous corporate engagement with climate policy. USCAP members sent the signal that for them the scientific debate was over and it was imperative to take action (interview 40). Internal negotiations on USCAP's policy position took place in a number of committees. Staffed by senior representatives of the alliance members, committees sometimes convened as often as every week.[16] These committees were cochaired by NGO and business representatives (interview 26). The executive committee, consisting of the chief executives of the member companies, held the agenda-setting power within USCAP. Hence, the coalition was very much CEO led.

Negotiations on the general policy approach led to consensus relatively quickly, with conflicts only arising over the details of the cap-and-trade scheme. According to participants in the creation of USCAP, the environmental groups pushed hard for cap and trade, even though some of the companies could have accepted a carbon tax (interview 40).[17] The decision on the regulatory approach was quickly settled in favor of emissions trading, while the real battle was over the rules of the trading scheme (Mouawad 2008). Issues of contention were the allocation method and price containment mechanisms. Big emitters were eager to ensure that emissions permits would be allocated for free. Environmental groups and less carbon-intensive businesses were in favor of auctioning. For instance, Lew Hay III, chair of FLP Group, a Florida electric utility, said that firms should pay for emissions rights. "There is just going to be a giant fight over the free allowances," he remarked (quoted in ibid.). Regarding price containment, the battle lines ran between industry and environmental groups (ibid.). Industry representatives argued for a ceiling on the price of emissions permits to keep the economic effect of carbon pricing at bay. Environmental groups feared that a ceiling might jeopardize the environmental effectiveness of a trading scheme. As in the EU, tensions arose between actors advocating a stringent regime and those asking for a lax regime. In July 2006, the chief executives of the coalition partners met to discuss the draft policy position. Participants reported that the executives asked the senior staff members negotiating the position to adopt bolder language (interviews 25 and 40).

USCAP's launch lent the proregulatory forces in U.S. business a strong voice and brand, making business conflict over climate policy apparent. For a significant portion of business, the voluntary approach was no longer tenable. It is interesting to see how the creation of this new collective forum served as a tipping point, causing a significant shift in corporate political strategies. Such bold collective action had effects on the landscape of business interests, which started to reconfigure after the USCAP was established. The ICCP (2007) closely aligned itself with USCAP by commenting positively on its launch and practically supporting USCAP's standpoint.

In March 2007, a broad coalition of financial actors released a call to action along similar lines. The group of more than sixty leading institutional investors (such as pension funds, financial service firms, and foundations) and a number of leading corporations (including BP, DuPont, and PG&E) was convened by Ceres, a coalition of institutional investors, along with the Investor Network on Climate Risk. The group asked the federal government to enact national policy that establishes a price on carbon through a cap-and-trade scheme. Furthermore, a number of sectoral and umbrella business organizations repositioned themselves on climate regulation. While none of these groups supported mandatory emissions controls, they acknowledged that they are to come. Expecting climate legislation, they outlined policy preferences or general policy principles that should be applied to any regulatory framework. In February 2007, the EEI was the first sectoral association to shift its climate policy. The "EEI Global Climate Change Principles" state that the association "supports federal action or legislation to reduce greenhouse gas emissions that . . . [inter alia] employs market mechanisms to secure cost-effective GHG reductions" (EEI 2007).

In contrast to the proregulatory and moderate industry voices, a number of associations continued to reject any form of mandatory climate regulation. Major actors in this group included API, Industrial Energy Consumers of America (IECA), and NAM.[18] These groups were upset by USCAP's call for action. For instance, soon after the launch of USCAP, Immelt received a call from John Wilder, chair of the electric utility TXU, expressing his concerns about the impact of potential carbon limits on the coal plants that TXU was planning (Kranhold 2007).

While a strong coalition has not emerged, opponents to a domestic cap-and-trade scheme mainly pursue two strategies: they portray the consequences of cap and trade as an economic disaster, and propose alternative abatement strategies, such as energy-efficiency improvements and technology development. The U.S. Chamber of Commerce, for instance, launched an ad that "shows a man cooking breakfast over candles in a cold, darkened house, then jogging to work on empty highways, asking: 'Is it really how Americans want to live?'" (cited in Mouawad 2008). NAM instead proposed to focus more on improving the energy efficiency of the economy than on setting up a cap-and-trade scheme. Keith W. McCoy, NAM's vice president for energy and resources policy, called the cap-and-trade approach a "fundamentally flawed" strategy (quoted in Garner 2008).

Unlike the proregulatory business faction, the antiregulatory business forces did not organize into a coalition. Their opposition remained diffuse. One lobbyist suggested that antiregulatory groups were too confident that legislation would not happen (interview 40). This mirrors the development in the EU, where the opposition only created an alliance because of the decision to develop the EU ETS. The simple correlation of the organization of a protrading coalition and the rise of cap and trade on the political agenda, on the one hand, and the lack of collective action in the antiregulatory camp, on the other, suggests that collective action matters as a source of political power. Anecdotal evidence indicates that antiregulatory groups overestimated their political clout by extrapolating from previous victories. This strategic miscalculation led them to not build a coalition. Hence, the campaign for cap and trade continued to gain political momentum.

The emergence of the protrading lobby coevolved with activities on climate legislation in Congress: the more active Congress was on the issue, the more vocal business was, and the more vocal business was, the more market-based climate bills gained momentum in Congress. Again, the diffusion of emissions trading is a case of multilevel leadership.

Congress Responds: Cap-and-Trade Bills, 2007–2009

Congressional activity on climate change reflected the changes in industry coalitions and state-level activities. The story of market-based climate bills in Congress is one of incremental progress and political momentum building up slowly. Between 1998 and 2002, the GCC was disintegrating, while the majority of U.S. business was supporting voluntary commitments on emissions reductions. Both Democrats and Republicans introduced a number of bills during this period that tried to address climate change by giving industry credits for early action to reduce GHG emissions, promoting voluntary emissions reduction commitments, and offering tax incentives for energy-efficiency measures (Layzer 2007). Yet political support for these proposals was weak, and none of them suggested mandatory emissions cuts. This changed in 2003, when Congress started to respond to increasing pressure from states, corporate leaders, and the environmental community by proposing market-based climate bills. From 2003 to 2006 (in both the 108th and 109th Congress), a few cap-and-trade proposals were brought to the table, but did not receive much support. The political developments and emergence of strong business support turned the tide. Since then, the momentum for federal cap-and-trade regulation has been building up, as policymakers consider legislation to be inevitable.

The 110th Congress saw the introduction of a great number of bipartisan market-based climate bills in both houses (Arimura et al. 2007).[19] This coincided with USCAP's launch and a number of calls from industry groups to enact a federal carbon-trading system in the United States. Some of the proposals were general, while others were detailed, with the latter including in particular the McCain-Lieberman, Bingaman-Specter, and Lieberman-Warner proposals. These three proposals received the most attention among policymakers, industry, and environmental groups. After the general elections of 2008, which resulted in a decisive Democratic majority in the 111th Congress, the House passed the first cap-and-trade bill, whereas antiregulatory business forces hindered legislation in the Senate.

110th Congress: The McCain-Lieberman, Bingaman-Specter, and Lieberman-Warner Bills

In January 2007, Senators John McCain and Joseph Lieberman reintroduced their proposal as the Climate Stewardship and Innovation Act, which gathered support from green groups in particular. On January 22, USCAP was launched in Washington, DC, providing strong momentum for market-based bills in Congress. The launch received considerable media attention. USCAP engaged with the authors of all major bills and vocally called on Congress to enact legislation. In a February 2007 hearing before the House Committee on Energy and Commerce, John Rice (2007, 37), GE's vice chair, said: "As the cornerstone of responsible regulation, a carbon cap and trade system will go a long way to further innovation, help the environment and improve the overall effectiveness of the world energy system."

Proregulatory advocacy received a significant boost in April 2007. The U.S. Supreme Court ruled by five to four that the EPA had the authority and duty to regulate GHG emissions from automobiles under the Clean Air Act (Scott 2007b). Observers suggested that the ruling would embolden congressional action on climate change. Moreover, the decision also mobilized industry actors that were not expecting emissions reduction mandates, because even in the absence of congressional action the EPA would have to regulate. Philip Clapp, president of the National Environmental Trust, noted that "this gives the next president of the United States the power to impose his or her own global emissions reduction program even if Congress doesn't act" by 2008. He added: "Every major industry is going to see that writing on the wall" (quoted in ibid.).

Against this backdrop, Senators Jeff Bingaman and Arlen Specter introduced the Low Carbon Economy Act in July 2007. Since summer 2005, Bingaman had been preparing his proposal, closely following the recommendations of the National Commission on Energy Policy. The commission is a bipartisan multistakeholder group whose seventeen commissioners come from academia, business, environmental advocacy, labor unions, and policymaking. At the end of 2004, the commission published the report *Ending the Energy Stalemate*, in which it proposed a cap-and-trade scheme including a safety valve for cost containment. The Bingaman-Specter bill adopted this element, making it popular among business constituencies.

Bingaman himself said he was hoping to strike a middle ground between the "sense of the Senate" resolutions, which were more symbolic than substantial, and the more stringent market-based climate bills that had been defeated in Congress (Scott 2005).

At the time of Bingaman's proposal, the debate on climate change had different contours than when the first market-based climate bills had been introduced. A cap-and-trade scheme was portrayed as delivering a win-win-win solution. It could reduce GHG emissions, stimulate a green economy, and reduce energy dependence. In January 2007, Bingaman said: "There's concern over global warming and greater discussion about moving to clean energy for our country and reducing dependence on foreign oil" (quoted in Cook 2007, 69). Against the background of the Iraq war, the goal of energy independence gave extra momentum to climate policy development. Though Bingaman was aiming for a middle ground, his bill was perceived to serve industry interests, while the McCain-Lieberman proposal was more aligned with the preferences of environmental advocacy groups.

In October 2007, Senators Lieberman and Mark Warner introduced the Climate Security Act, which in many ways reflected the compromise solution between business and environmental groups.[20] The plan was therefore touted by some as potential breakthrough legislation (Ware 2007b). The competition between climate bills was leading to a convergence, increasing the likelihood of adoption in Congress (Arimura 2007). Indeed, many companies and environmental groups expressed support for the bill (Lieberman 2007). While USCAP (2007b) had been hesitant to endorse any climate change bill up to that point, it sent a letter to Senators Lieberman and Warner in October 2007, urging the respective committees "to report out a bill that can be taken up and passed by the Senate in this Congress." The letter stressed that USCAP believed that the Climate Security Act addressed many of its recommendations.

Despite collective support for the Lieberman-Warner bill, some USCAP members, such as the carbon-intensive utility Duke Energy, were strongly opposed to the bill, arguing that it put too high a burden on coal-based energy. Rogers, Duke Energy's chief executive, said: "Only the mafia could create an organization that would skim money off the top the way this legislation would skim money off the top" (quoted in Eilperin and Mufson 2008). The criticism referred in particular to the allocation method. The Lieberman-Warner bill, like virtually all market-based climate bills in Congress, allocated only 75 percent of allowances for free, while 25 percent would be auctioned. The share of auctioned permits increased over time until all allowances would be auctioned. Rogers considered this a financial disaster for Duke, as he estimated that the company would have to spend about $2 billion in the first year (Thompson 2008). Moreover, according to the bill, the revenues from auctioning would be given to consumers as tax refunds and used to reduce the state deficit. Rogers's criticism was that the revenues would not be channeled primarily toward investment in clean-energy technologies. When the bill was brought up for a vote, Duke Energy reached out to its customers to ask them to lobby senators to vote against it (Eilperin and Mufson 2008). Though these tensions did not cause the coalition to break apart, they reflected the distributional struggles evolving when it comes to designing the rules of cap-and-trade schemes.

Under the leadership of Senator Barbara Boxer of California, the Environment and Public Works Committee approved the bill by a vote of eleven to eight in December 2005. For the first time in U.S. climate politics, a bill that included mandatory emissions limits passed a senate committee. In June 2008, the bill came to a floor vote in the Senate, with forty-eight senators voting in favor of it and thirty-six rejecting it.[21] Another six senators who were absent wrote letters saying that they would have voted for the bill. Yet sixty votes were needed for the bill to pass the Senate (Zabarenko 2008).

Despite the defeat of the legislative proposal, supporters were enthusiastic about the vote, as a mandatory climate bill had never received as much support before. A coalition of green groups declared: "Today's vote set the stage for a new president and Congress to enact strong legislation that will more effectively build a clean energy economy and prevent the worst consequences of global warming" (cited in ibid.). Other supporters included thirteen labor unions, the U.S. Conference of Mayors, and a number of religious groups (Eilperin and Mufson 2008). Nevertheless, the bill also drew strong criticism from energy-intensive industries, carbon-intensive energy companies, and Republican senators. John Engler, NAM's president, said that the bill was tantamount to "economic disarmament" (quoted in Mouawad 2008). The economic significance of the bill is underlined by the size of the market that it would have created. The research firm Energy Finance estimated that if the Lieberman-Warner bill was in fact implemented, the market would reach $1 trillion by 2020 (Greening of Wall Street 2008).

Reflecting on the Senate vote on the Lieberman-Warner bill, Senator Boxer observed that "in America change doesn't happen overnight, it takes time to turn the ship of state. . . . This is coming" (quoted in Zabarenko 2008). The case of the Lieberman-Warner bill shows both the growing influence of the protrading coalition and the limits of its influence. USCAP was well organized and represented a large percentage of the U.S. economy. The protrading coalition helped to increase the pool of supporters for a market-based climate bill in Congress, but at the same time, it could not yet mobilize sufficient support for a bill's passage.

111th Congress: The Waxman-Markey Bill

The battle over federal cap-and-trade bills reached a new level of intensity in the 111th Congress, which had a strong Democratic majority. Congresspersons Henry Waxman and Edward Markey introduced the American Clean Energy and Security Act in the House of Representatives on March 31, 2009. The legislation was introduced in view of COP 15 in Copenhagen in December 2009, which was meant to produce an agreement on a post-2012 international climate regime. The political rationale was that the United States should agree on a domestic emissions reduction target before an international agreement would be negotiated. USCAP (2009) publicly welcomed the proposal as a "strong starting point." The bill incorporated many of USCAP's recommendations—which it had published in January 2009—for the design of a cap-and-trade scheme. These suggestions included an economy-wide cap-and-trade scheme and provisions for cost containment. The Waxman-Markey bill provided favorable conditions, especially for electric utilities (Broder and Mouawad 2009), mostly because they were at the negotiating table. In contrast, the oil and gas companies, which had not engaged with legislators but had opposed the bill, found that the legislative proposal was especially unfavorable to their industry. On June 26, the House passed the act by a 219–212 majority.

In the aftermath of this event, conflict erupted within the U.S. business economy once more. The U.S. Chamber of Commerce, API, NAM, the American Coalition for Clean Coal Electricity, and the American Farm Bureau Federation were at the forefront of the opposition to the bill (Mouawad 2009). This led to a number of companies defecting from these associations, including electric utilities such as Exelon, PG&E, and PNM. Companies left the U.S. Chamber of Commerce, in particular, because of its plan to run a campaign that questioned the science of climate change. Peter Darbee, the chief executive of PG&E, said, "We find it dismaying that the Chamber neglects the indisputable fact that a decisive majority of experts have said the data on global warming are compelling and point to a threat that cannot be ignored" (quoted in Fahrenthold 2009). Questioning the science would have meant a return to the corporate political strategies employed in the 1990s.

This visible shift of companies from the antiregulatory coalition to the protrading coalition had an impact on policymakers. Congressperson Jay Inslee commented, “There will be significant vibrations from this. It’s a bit of an earthquake” (quoted in Krauss and Galbraith 2009). Bingaman, a Democrat and chair of the Energy and Natural Resources Committee, said he did not know what impact the public announcements of firms had on individual senators. “But I do think it’s a sign at least some in the business community are anxious to see us provide some leadership on climate change” (quoted in ibid.).

With the house bill having passed, it was the Senate’s turn to produce legislation. The Senate Environment and Public Works Committee, chaired by Boxer, passed the Clean Energy Jobs and American Power Act of 2009, on November 5. The bill drew heavily from the Waxman-Markey bill, also proposing a federal cap-and-trade scheme, but the Senate was preoccupied with health care reform, and the senators from coal states were reluctant to move quickly on climate legislation. Heavily coal-based utilities proved to still have considerable clout over how the senators from their home states would cast their vote. In the end, Congress did not pass a comprehensive climate bill before the Copenhagen climate summit. Nonetheless, President Barack Obama announced a national emissions reduction target for the United States of 17 percent below the 2005 levels by 2020. This was along the lines of the targets put forward in the House and Senate bills.

By the end of 2009, the United States had an operational regional cap-and-trade scheme, and the idea of a federal cap-and-trade scheme was firmly embedded in the climate agenda. While the protrading coalition had managed to maintain the momentum for carbon trading, the antiregulatory business forces had succeeded in preventing the passage of climate legislation in the Senate and they played out their influence in the coal states. At the time of this writing, the Senate has abandoned its attempt to pass climate legislation in 2010. It remains to be seen how the power game between pro- and antiregulatory business forces will be decided in federal U.S. climate politics. Meanwhile, U.S. states are continuing to develop regional trading schemes, which could potentially be linked to schemes abroad such as the EU ETS.

Conclusion

The reemergence of carbon trading on the political agenda in the United States went along with the creation of a classic baptist-and-bootlegger coalition—USCAP. USCAP played an important role in accelerating the political momentum for carbon trading in the United States, but it was embedded in a broader campaign for emissions trading schemes, which included state policymakers. In this process of multilevel leadership, USCAP's influence is best described as critically accelerating a preexisting momentum, which propelled the development of cap-and-trade bills in Congress. It enabled policymakers to advance mandatory climate policy.

This is a recurring theme throughout the three case studies: business did not change the policy preferences of states but instead created the enabling conditions that allowed governments to move forward with emissions trading as a form of mandatory climate policy. Firms could not prevent mandatory climate regulation but instead could decisively influence the regulatory style. In both the EU and the United States, business groups had exercised de facto veto power with regard to carbon-energy taxes. When they recognized that they could not prevent mandatory emissions cuts, however, they settled on emissions trading as the compromise, which they successfully pushed through. The question, of course, is why USCAP succeeded in moving the idea of a nationwide cap-and-trade scheme firmly on the U.S. political agenda at the time it actually did. After all, an earlier attempt by the Partnership for Climate Action had failed. In many respects, political opportunities played a pivotal role in the sudden influence of the protrading coalition in the United States.

In the U.S. case, there was no immediate policy crisis in the policy stream that created a political opportunity for protrading advocacy (see also table 6.1). Yet Hurricane Katrina presented a moderately important crisis in the problem stream, as it was widely framed as a result of global climate change. More important, though, the ratification of the Kyoto Protocol and the emergence of the EU ETS created a normative momentum for mandatory market-based climate policy, which increased the pressure for domestic climate action in the United States. The normative fit of carbon trading with the liberal norms of the U.S. economy also offered an opportunity. It was a truism in the U.S. business community that if companies were ever to advocate binding emissions cuts, emissions trading would be the way to go (interviews 33 and 38). Opposition to climate policy was never directed at the market mechanisms per se, as it had been in Germany, but rather at the hard caps that come with them. With the exception of the environmental justice movement, every major political actor in the U.S. had a preference for emissions trading as opposed to carbon taxes or technology standard, given the positive experience with SO_2 trading under the Clean Air Act. This was true for state governments as much as industry.

Table 6.1
The sources of influence of the protrading coalition in the United States

Source of influence	**Impact**	**Evidence**
Political opportunities		
Policy crisis	Moderate	Hurricane Katrina presented a crisis in the problem stream, but it was only one among several political opportunities that aligned
Norms	High	The entry into force-of the Kyoto Protocol created the international normative momentum for climate action
Coalition resources		
Financial resources	Low	USCAP asked for membership fees, but these fees only played a role for funding limited organizational capacity
Legitimacy	High	Legitimacy was the key resource in prorading advocacy; the comparative legitimacy advantage stemmed particularly from the fact that two antagonistic actors (business and NGOs) cooperated and from the representativeness of USCAP
Coalition strategies		
Mobilizing state ally	High	The U.S. states and sponsors in Congress were important state allies
Playing multilevel games	High	Advocates targeted state-level and federal politics

Regarding the coalition's resources, legitimacy played a greater role as a source of influence than funding. Though both NGO and business participants paid membership fees, neither group saw the pooling of campaign funding as a major motive for cooperation and advocacy success. Moreover, USCAP did not have an advantage in funding over the loose but well-funded antiregulatory coalition. Funding thus does not explain why the protrading coalition prevailed over the antiregulatory one. USCAP's comparative legitimacy advantage builds on two dimensions: by gathering companies from a variety of economic sectors, and more significantly, by representing both environmental and economic interests. Usually policymakers face the task of squaring a number of vested interests in policymaking. Baptist-and-bootlegger coalitions perform this task within the alliance by prenegotiating a compromise, which they then present to government. By generating a compromise solution that was supported by a range of different actors, USCAP delivered both interest aggregation and the pooling of legitimacy resources. The ability of an advocacy group to deliver a compromise solution to policymakers hinges directly on its ability to organize collective action between adversarial actors and conflicting interests.

The protrading coalition relied heavily on U.S. states and members of Congress in setting the agenda in favor of a domestic cap-and-trade scheme. Climate leadership by California and states in the Northeast was an essential prerequisite to federal politicians considering mandatory climate policy at all. This shows a striking parallel to what happened in the EU when the United Kingdom and Denmark pioneered emissions trading. Subfederal actors leveraged action at a higher polity level. Governors in all states taking climate action were in favor of emissions trading, which made them natural allies for companies that tried to ensure the use of market-based climate policy. This was of particular strategic significance as state policies were likely to set the agenda for national climate politics. At the federal level, USCAP could increasingly rely on Congress to support climate policy, especially with a Democratic majority. Congress was an ally for industry, as influential members of both parties shared a preference for market mechanisms. The reasons for their policy preference are the positive experience with the SO_2 trading program and the prospect of a large stream of revenues from auctioning permits. Though key government actors were in favor of emissions trading, business had to organize to demonstrate business support and be able to defend its specific design preferences for a cap-and-trade system—that is, to mitigate the regulatory risk.

The question here, of course, is what came first: state action or business action. Did business organize collective action in response to state action, or did states respond to business collective action? As a classic chicken-and-egg problem, this issue cannot be resolved completely. Business advocacy and state activities coevolved, both in how state action mobilized business and how business advocacy triggered state action. It seems that business sensed a higher willingness to act among policymakers and decided to get ahead of the curve by pushing for a specific regulatory approach. The creation of a new collective voice with strong branding as a cross-sectoral alliance in turn gave weight to the cap-and-trade agenda, spurring legislative action.

Finally, the case demonstrates that successful coalitions play political games across a number of levels. The protrading campaign focused first on the state level before it targeted the federal one. It thus leveraged pressure from below. The fact that the protrading coalition addressed different venues at different times shows that the distribution of political authority in climate governance demands campaigns to target multiple levels of political decision making in order to affect a large-scale change of agendas. BP, DuPont, and others got involved in California to set an important precedent for the federal climate politics agenda in the United States. The same is true for RGGI. Actors involved in the different local political processes remained connected transnationally for the purposes of information exchange and—to some degree—the coordination of political strategies. With carbon trading firmly anchored on the U.S. climate agenda, the project of international GHG emissions trading had been firmly embedded in the transatlantic and global agenda. At the same time, the project of a globally integrated carbon market was still in its infancy.

7

Business and the Rise of Market-Based Climate Governance

In the decade following Kyoto, a new currency emerged in the global political economy: carbon credits. The diffusion of carbon trading across the Organization for Economic Cooperation and Development world and major developing countries occurred surprisingly quickly. It represents one of the most significant and most recent developments in the broader shift from command-and-control regulation to market-based forms in global environmental politics. Like no other environmental policy, carbon trading moves environmental policy into the heart of the world economy: the energy and financial systems. While to date carbon markets remain highly fragmented and a global reference price does not yet exist, they are likely to grow over the coming decade. What is more, environmental markets will most likely emerge in other issue areas in the future.

This book set out to explore the role that business played in the rapid rise of emissions trading in global climate politics. Making the case for a neopluralist, coalition-centered reading of the emergence of carbon markets, I have argued that a transnational protrading business coalition was instrumental in enabling governments to go forward with mandatory yet market-based climate policy. The protrading coalition's influence has been studied in the cases of the Kyoto Protocol, the EU ETS, and the move in the United States toward cap-and-trade schemes. These three events are widely perceived to be the milestones in the globalization of carbon trading. The center of political action on carbon trading shifted from the international stage to the EU and then to the United States.

In the following, I will present the findings and draw conclusions. I first summarize the business-centered reading of the rise of carbon trading with a focus on the corporate campaign. The empirical narrative is followed by a discussion of the sources of the influence of transnational business coalitions against the backdrop of the empirical findings. Thereafter, the chapter debates the contending explanatory approaches set out in the introduction in light of the empirical findings. I will show that alternative theoretical approaches leave a number of established facts unexplained, while complementing the explanation offered here in some aspects. I will then discuss the findings in the context of the debate on the role of business in the shift toward market-based environmental governance. Finally, I consider the implications of the findings for research on global environmental politics as well as for policymakers.

Business and the Making of Carbon Markets

From a neopluralist, coalition-centered perspective, emissions trading emerged as the policy instrument of choice because a transnational coalition of mostly firms and a few environmental groups promoted it as a climate compromise (see also Meckling 2011). In this process, the coalition and its state allies outcompeted the antiregulatory business coalition, which preferred no emissions controls at all, on the one hand, and an NGO network, which called for classic command-and-control policies, on the other hand (see table 7.1). Though coming from opposite sides of the divide, both of these coalitions saw environmental protection and economic growth as being at odds with each other. Business therefore tried to avoid emissions reduction mandates, while environmental groups that had not bought into the emissions trading compromise advocated state-driven command-and-control policies.

The Kyoto Protocol: Internationalizing a U.S. Regulatory Approach

In the first phase of industry engagement with climate policy, the GCC and a few allying organizations, including the API and the ICC, organized transnational and transatlantic industry opposition to climate regulation, with U.S. oil companies being instrumental in the process. Until the mid-1990s, this antiregulatory business coalition was highly successful in averting international and domestic emissions reduction obligations. The GCC mainly pursued two strategies to influence climate policy in this early phase. It attempted to discredit climate science. Once the scientific consensus on climate change became less questionable, though, the GCC framed its opposition to climate regulations as a matter of economic costs and a question of the participation of developing countries.

The influence of the transatlantic business opposition to climate regulation is partly accountable for a number of policy outcomes at the international and domestic levels. First, the UNFCCC did not contain any binding emissions reduction targets and timetables. Second, in the aftermath of the Rio Earth Summit, carbon taxes were proposed in both Europe and the

Table 7.1
The protrading coalition in comparison with its competitors

	Pro-command-and-control NGO coalition	**Pro-trading NGO-business coalition**	**Antiregulatory business coalition**
Campaign period	1990–1997	1996–present	1990–present
Policy preference	Command-and-control policies	Global emissions trading	No emissions controls; from 2000 on voluntary climate policy
Normative frame	Economy versus environment	Economy and environment	Economy versus environment
Material interests	Reduce GHG emissions	Reduce GHG emissions; minimize compliance costs; create and expand markets	Avoid compliance costs
Key actors	CAN	BP, Shell, ED, IETA, DuPont, ICCP, and WBCSD	GCC, API, BDI, ICC, IECA, ExxonMobil, and NAM
Normative change	Need for mandatory emissions cuts internationally accepted	Market-based climate policy recognized as the default solution	Awareness of the cost of climate policy
Policy change	Targets-and-timetables approach in Kyoto Protocol	Cap-and-trade schemes in the EU, the United States, and other countries	Byrd-Hagel resolution; U.S. withdrawal from Kyoto Protocol

United States. Industry groups successfully fought those proposals. Third, U.S. industry groups were the driving force behind the Byrd-Hagel resolution, which was passed by the U.S. Senate in 1997 just before COP 3 in Kyoto. The Senate supposedly would not agree to an international treaty that did not include binding emissions reduction targets for developing countries. Thus, industry managed to undermine domestic support for the administration's foreign policy on climate change.

The balance of power between environmental and business interests began to shift when the Berlin Mandate was agreed to at COP 1 in 1995. The mandate said that governments should adopt quantified emissions reduction commitments by 1997. It is in this context of international and subsequently domestic policy developments that a rift on climate policy emerged in the transatlantic business community. European oil major BP and the green group Environmental Defense spearheaded a new political strategy among business and environmental groups that focused on the promotion of market-based climate policy. Market mechanisms appeared as the compromise solution between industry's reluctance to accept any kind of mandatory emissions targets and the environmental community's preference for command-and-control policies. Emissions trading was portrayed as the third way between the policy concepts of the political Right and Left, which allowed flexible and cost-effective implementation of emissions reduction targets. The new emerging NGO-business coalition in support of market-based climate policy weakened the antiregulatory business coalition and the pro-policies-and-measures environmental coalition.

In the run-up to Kyoto, the protrading coalition supported the Clinton administration's foreign policy on market-based climate policy and helped to mobilize support for market mechanisms among negotiating parties. In particular, the coalition worked to persuade European governments to accept market mechanisms, as governments, businesses, and NGOs in continental Europe had been highly skeptical of tradable permits. The inclusion of flexible mechanisms in the protocol reflected the preference of the protrading coalition. In the conference's aftermath, the transnational protrading coalition started to institutionalize as new organizations were created, including the Pew Center on Global Climate Change and the IETA. Once emissions trading was a constitutive part of the international climate policy framework, the political focus shifted to the ratification and implementation of the protocol at the national level, especially within the key entities such as the EU and the United States. The EU took the lead, and the United States followed with a certain time lag.

The EU: From Foe to Friend of Carbon Trading

Despite its initial opposition to tradable permits, the EU designed and implemented the first cross-border emissions trading scheme. The political momentum to go ahead with emissions trading in Europe grew from the bottom-up, starting in the United Kingdom. BP, the central broker of the transnational protrading coalition, led the process by setting up a global in-house trading scheme with the help of Environmental Defense. It thus demonstrated both the feasibility of trading schemes and the political will to support a market-based European climate policy. Under the influence of BP, UK oil and power companies saw an opportunity to put emissions trading firmly on the European agenda if the United Kingdom pioneered a trading scheme. They therefore set up the UK Emissions Trading Group with the support of the UK government. An advocacy coalition at heart, the UK ETG developed the UK ETS. The UK ETS and corporate leaders on emissions trading in general were very much driven by an antitaxation agenda, and as such, were mainly pursuing a proregulatory risk management strategy.

The pioneering work in the United Kingdom subsequently spurred action at the EU level, inter alia, to prevent regulatory fragmentation within the EU. While British oil majors BP, Shell, and Eurelectric, the association of the European power industry, were at the core of the protrading coalition, the European Commission became a powerful driver of an EU-wide scheme in its own right. The commission had a number of reasons for supporting emissions trading in Europe. First, the European Commission knew that a carbon tax was deemed to fail, as it had in the early 1990s because of business opposition. Business, in turn, was aware of the political will to implement mandatory climate regulation in the EU. Second, the development of the UK and Danish trading schemes spurred fears of regulatory fragmentation in the EU, which could have undermined integration achievements with regard to the internal market and environmental policy. Third, officials were critically aware of the fact that emissions trading was the price the EU had to pay to get the United States to ratify the Kyoto Protocol. Changing course on market-based climate policy thus improved the chances for the reengagement of the United States.

The protrading coalition did not remain unchallenged. German industry and energy-intensive manufacturing industries were particularly opposed to a European trading scheme. Unlike in the United States, environmental groups were not among the protrading advocates but were instead on the fence. They became watchdogs for the environmental effectiveness of the scheme only at a relatively late stage in the process. After the EU ETS had entered its implementation phase in early 2005, the coalition supporting it became increasingly at risk of fragmenting, as business was divided over the stringency of the system. The financial services industry started to advocate for a more stringent EU ETS, which the big emitters from the energy industry and energy-intensive manufacturing industries tried to avoid. At the same time, a number of business groups began to lobby for the expansion and integration of carbon markets by targeting the negotiations for a future international climate regime that would succeed the Kyoto Protocol. With the EU setting the pace on carbon trading, it was not long until the United States jumped on the international bandwagon of carbon trading.

The United States: Catching up with Carbon Trading

After the Kyoto Protocol agreement, the United States implemented voluntary climate policies for a number of years. During the Clinton administration, the Senate was the major hurdle for any mandatory climate policy. After the Bush administration withdrew the United States from the Kyoto Protocol in 2001, the executive and legislative branches were aligned regarding voluntary climate policy. Industry was able to mitigate public pressure for climate action while avoiding mandatory emissions cuts.

Yet business and state activities on the ground began to put emissions trading back on the U.S. agenda. In 2003, the CCX, a private initiative, established the first GHG emissions trading scheme in the United States. In the same year, a number of states in the northeastern United States set out to develop a regional emissions trading for the power sector—what would become the RGGI. In 2006, California passed the Global Warming Solutions Act, which kicked off a process to develop an economy-wide state-level emissions trading scheme. With business and U.S. states moving forward with mandatory and market-based climate policy, pressure to enact mandatory emissions cuts at the national level increased.

When key political parameters changed in national politics, a window of opportunity opened for a mandatory federal climate policy. Congress was increasingly responding to the pressure from below by developing cap-and-trade bills. The Democrats' win in the 2006 midterm elections increased the momentum for climate legislation in both the House and Senate. In addition, the international momentum for mandatory climate policy was growing after the Kyoto Protocol had entered into force in 2005. Unlike in any previous phase of U.S. climate politics, from 2004 on climate regulation was looming over business as a realistic threat. This ultimately led to a deep split within U.S. industry over political strategy.

A number of climate leaders such as Cinergy and GE were demanding federal climate legislation. This emerging business conflict ultimately led to the creation of USCAP in January 2007. A protrading NGO-business coalition, USCAP advocated a federal cap-and-trade scheme for the United States. The mainstream of business was now clearly divided over the course of U.S. climate policy. With a strong political-economic coalition emerging, legislative activities on market-based climate policy picked up, with the first cap-and-trade bill coming to a Senate floor vote in 2008. By 2010, the first regional trading scheme was operational, others were being developed, and the House of Representatives had passed the first market-based climate bill. The passage of a federal bill by the entire Congress was still pending.

While the EU and the United States have been the prominent industrialized players in global climate politics, other countries are following suit

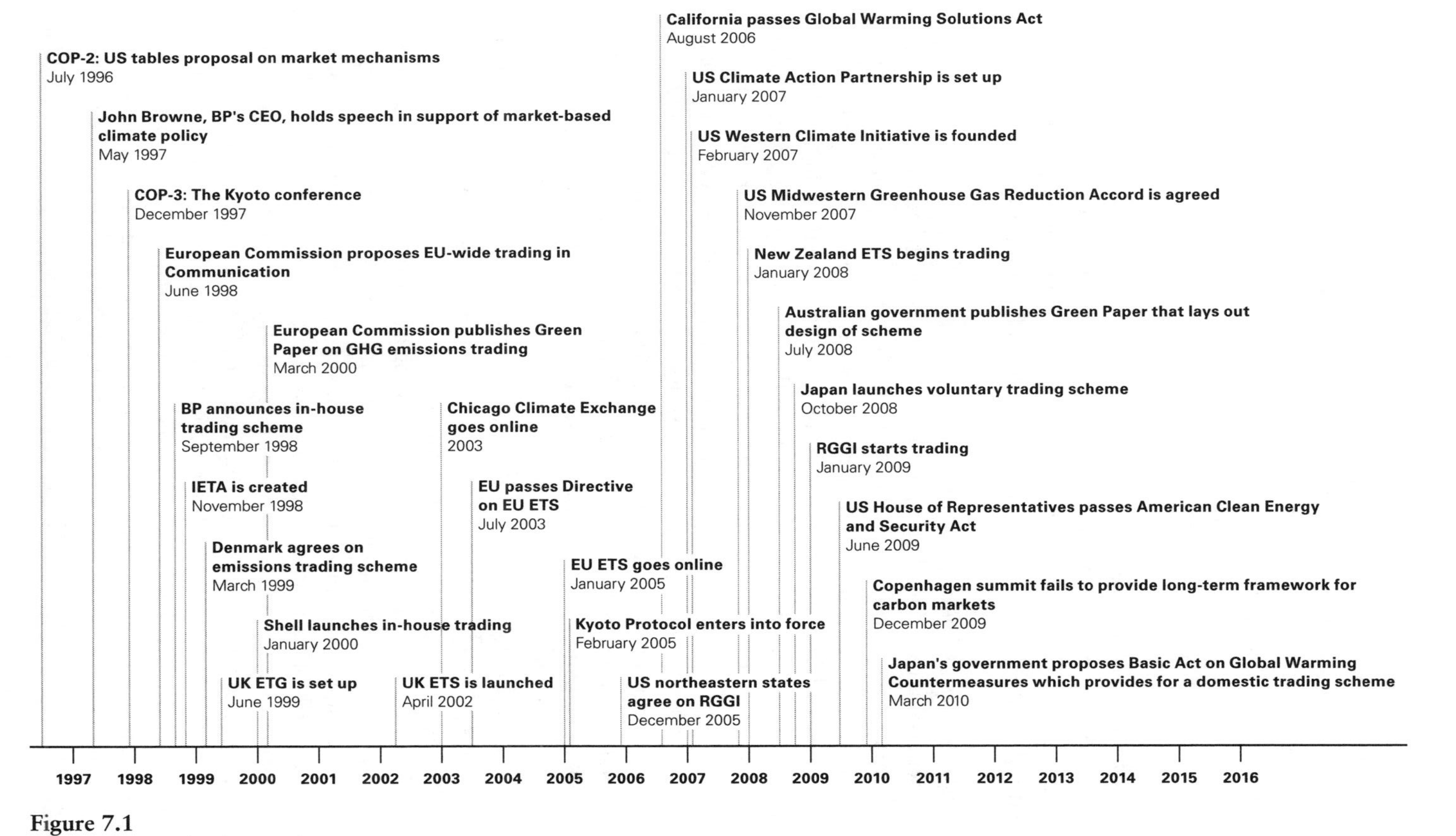

Figure 7.1
The global rise of carbon trading

with emissions trading schemes (see figure 7.1). Anecdotal evidence suggests that a business-centered, protrading coalition also played a critical role in Australia. In 2006, the Business Roundtable on Climate Change recommended that Australia implement a market-based carbon pricing mechanism (Griffin 2006). In July 2008, the Australian government announced its plan to implement an economy-wide cap-and-trade scheme by 2010 (Taylor and Thornhill 2008).

Key Empirical Findings

The story of the rise of carbon trading holds a number of interesting empirical insights with regard to the type and origin of corporate policy entrepreneurs along with the pattern of the diffusion process.

Big Emitters as Policy Entrepreneurs

Big emitters from the oil and power industries constituted the core membership of the protrading coalition. In all three cases, advocacy for emissions trading was primarily understood as a risk management strategy. It has been widely assumed in the business community that the regulatory risk of emissions trading is lower than that of a carbon tax. Hence, the agenda behind the rise of emissions trading is very much an antitaxation one. The market-creating (carbon markets) and market-expanding effect (low-carbon technology markets) of emissions trading only emerged as a motive for protrading advocacy once the EU ETS had been implemented, bringing new actors to the table such as financial services and large technology companies. As will be discussed later, this finding is not commensurate with neo-Gramscian notions that global finance is the structural force driving the globalization of emissions trading markets (Paterson and Newell 2008). The early advocates of carbon trading were not financial services providers but rather big emitters.

While big emitters saw emissions trading as a way to hedge regulatory risk, government actors and environmental groups had their own motives for supporting the instrument. The reasons why government actors backed carbon trading include a positive institutional memory regarding emissions trading (the U.S. administration), the prospect of a significant revenue stream from auctioning permits (the U.S. Congress), and strategic concerns related to an international agreement on a climate treaty and EU-internal power dynamics (the European Commission). As for the NGO community, business-oriented environmental groups pushed the emissions trading agenda because they were able to mobilize business support for it as opposed to alternative mandatory climate policies. Moreover, carbon trading provides certainty about the quantity of emissions reductions, which lets it score high on environmental effectiveness. Altogether, while for the big emitters risk management was the predominant motive for protrading advocacy, the overall coalition was propelled by various motives. There is no simple common denominator that explains which kinds of environmental policy instruments can find broad backing across constituencies. Later in this chapter, I will discuss why it would be misleading to assume that market-based environmental policy in general can mobilize broad political-economic coalitions.

An Anglo-American Advocacy Coalition

The key protrading advocates are all based in Anglo-American countries—that is, the United Kingdom and the United States. This reflects the fact that emissions trading is very much rooted in the institutions, norms, and business-government relations of liberal market economies. In this respect, the diffusion of emissions trading mirrors and reinforces the strength of the liberal market economy model as opposed to that of coordinated market economies. It is an interesting observation that multinational corporations acted as transmission belts for a regulatory policy instrument, which initially did not have many friends in the coordinated market economies of mainland Europe. This raises question as to whether transnational business coalitions represent a critical channel for the diffusion of liberal policies more generally (cf. Simmons, Dobbin, and Garrett 2008).

Market Creation: From Private Demonstration to Public Trading Schemes

In both cases—the EU and the United States—carbon trading emerged in a bottom-up fashion. In a first step, private projects pioneered carbon trading, thus demonstrating its feasibility and the political will of business to support market-based climate policy (see table 7.2). In the EU,

Table 7.2
From private experiments to public trading schemes

	EU	United States
Private pilots	BP, Shell, and Eurelectric	CCX
Subfederal pilots	UK ETS and Danish scheme	RGGI and California
Federal scheme	EU ETS	Cap-and-trade bills in Congress

BP's in-house trading was the first scheme, while in the United States the CCX set a precedent. Public trading schemes at the subfederal level followed. These acted as strategic levers for the creation of federal trading schemes. The UK ETS and RGGI are cases in point. The main purpose of the two schemes was to trigger market creation at a higher polity level. In a third step, a federal scheme was developed that superseded the subfederal scheme. The EU ETS is up and running in the EU, while the U.S. Congress is moving toward passing legislation for an economy-wide trading scheme. In short, after the initial internationalization of the idea of emissions trading through the Kyoto Protocol, carbon trading developed in a bottom-up process with business actors taking the lead by constructing a case for emissions trading (cf. Victor, House, and Joy 2005).

The bottom-up emergence of carbon markets suggests striking parallels with the emergence of the international trading system for goods and services as well as the EU's internal market (cf. Victor and House 2004; Victor 2007). Both of these trading systems proceeded through different degrees of integration—from the General Agreement on Tariffs and Trade to the World Trade Organization, and from the European Economic Community to the EU. While the current carbon market is highly fragmented, its bottom-up emergence and successive integration suggests that it is following a classic pattern of market emergence.

To sum up, this study has offered a business-centered, coalition-oriented reading of the rise of emissions trading. While industry could not prevent climate regulation, it critically affected the style of regulation in favor of emissions trading. The protrading coalition acted as the political-economic engine of the spread of emissions trading.

The Influence of Transnational Business Coalitions Revisited

Speaking to the debate on the role of business in global environmental politics, this book has argued for a concept of transnational business coalitions to complement the study of business conflict. Business conflict limits the influence of business, while business coalitions enhance it. Business cooperation represents an organizational source of power, which allows firms to pool power resources and leverage collective political strategies. The transnational coalition perspective has proven useful in detecting broad patterns of collective action and interest group competition across time and political levels. While the member base and communication between protrading advocates were transnational in nature, the members actively involved in any given political process varied slightly from instant to instant. The transnational coalition essentially crystallized in more contextual coalitions such as the UK ETG. We observed varying degrees of cooperation in terms of the density of exchange and duration ranging from an ad hoc instrumental coalition prior to Kyoto to a highly institutionalized campaign coalition such as USCAP.

In the following, I will revisit the role of international political opportunities, coalition characteristics, and coalition strategies for the influence of transnational business coalitions. Policy crises, normative fit, and financial resources played a moderate role as sources of influence, whereas legitimacy along with the ability to mobilize state allies and play multilevel games were all key. The protrading coalition's agenda was successful because powerful and like-minded state allies within a number of prominent powers supported it, and because it offered a compromise solution that was supported by business groups as well as NGOs. Finally, a number of additional factors could be identified from the case studies—all deserving further investigation. These include in particular the institutional memory of government actors as a political opportunity structure and venue shopping as an essential campaign strategy.

International Opportunities and Constraints: Crises and Norms

The political environment of firms creates opportunities and constraints for advocacy campaigns. Both policy crises and normative fit have proven to be conducive to the success of advocacy. Yet they do not represent the necessary conditions.

Policy Crises: Calling for New Policy Ideas

Policy crises represent opportunities for advocacy because they question conventional wisdom and call for new policy solutions. This is true for crises in both the policy and problem stream. The stalemate between environmental and economic interests in the early phase of climate politics created an opening for a new policy idea that would help to overcome the policy crisis. The proposal of mandatory yet market-based climate policy in the run-up to the Kyoto conference offered a way out of the stalemate. While this crisis lay in the policy stream, the case of the United States demonstrated how a crisis in the problem stream could create advocacy opportunities. Hurricane Katrina was one among a number of factors that created political momentum for mandatory climate legislation. The media widely framed the hurricane as a result of global climate change, which stressed the severity of the problem to the public and policymakers, thereby implying the need for a policy solution, which then could be exploited by proregulatory forces.

Against this backdrop, two specific lessons can be drawn for the study of policy crises as political opportunity structures in global environmental politics. First, close attention should be paid to the polarization of environmental and economic interests, which can create space for "third-way" solutions. Second, analysts should consider how environmental catastrophes and new environmental information create enabling or constraining conditions for advocacy campaigns. The case studies together demonstrate that policy crises in both the policy and problem stream create opportunities for advocacy, but that they are only one among several factors in the political environment. Policy crises created opportunities in the Kyoto and U.S. cases, whereas the EU case did not feature a policy crisis. They are not necessary conditions for successful advocacy.

Norms: Like Likes Like

The fit between a policy idea and domestic and/or international normative structures has proven to be conducive to the success of advocacy. Regarding domestic norms, the actual pattern of successful protrading advocacy followed the proposition that emissions trading is an easy sale in liberal market economies. Prior to Kyoto, British and U.S. firms were at the core of the protrading coalition lobbying the U.S. administration. In the case of the EU, BP, Shell, and electric utilities were the key drivers behind a Europewide coalition, which resonated well with the European Commission. Firms from continental Europe initially did not advocate the instrument, as they favored voluntary agreements. In both the United Kingdom and the United States, protrading advocates could relatively easily mobilize support from government and other political actors for market-based climate policy once the momentum for mandatory climate policy was strong. The normative fit in this respect provides an explanation for the emergence of state allies, which will be discussed in more detail later in this chapter. Not only domestic norms but also international ones have proven to present political opportunities. In the case of the EU, protrading advocates could refer to the market mechanisms of the Kyoto Protocol, which created normative support for their domestic advocacy campaign. Activists in the United States, in a similar vein, could draw on the Kyoto Protocol's ratification to demonstrate the international momentum for mandatory climate action. This in turn raised the heat for domestic action in the United States.

Yet a good normative fit is not a necessary condition for successful advocacy campaigns. The EU case study has shown how German firms and the German government supported emissions trading after strong initial resistance. For instance, German power companies and Électricité de France became ardent supporters of emissions trading once they had learned about the scheme's material benefits. The EU ETS allowed power companies to reap windfall profits and gave a boost to nuclear energy. Hence, institutional and normative barriers to a market-based instrument could be overcome through learning about the costs and benefits of a particular policy. This led to firms from continental Europe buying in to the transnational protrading coalition. This observation aligns with arguments made by Beth Simmons and Zachary Elkins (2004) on the diffusion of policies by learning. While their contention applies to states and the globalization of liberal economic policies, it can well be transferred to the context of business and global environmental policies. The mechanism of learning is the same. Thus, transnational campaigns may experience domestic norms as obstacles, but these may erode over time as domestic actors learn about the policy idea.[1]

To conclude, policy crises and norms are constitutive elements of the political environment, which enable or constrain advocacy campaigns. The case studies have shown that other environmental factors matter, too. For example, the positive institutional memory of the U.S. administration regarding its experience with SO_2 trading made it much easier for protrading advocates to mobilize allies within government. A number of elements of the political environment of nonstate actors therefore can enable or constrain advocacy. Policy crises and norms have proven to be important environmental factors, but not indispensable factors for successful advocacy.

Coalition Resources: Financial Resources and Legitimacy

The influence of transnational coalitions depends on not only conducive political conditions but also a coalition's resource equipment. Financial resources and legitimacy are key sources of instrumental and discursive power. While the case studies show that legitimacy was pivotal to advocacy success, financial resources appear to have played a more limited role.

Funding: Material Underpinnings of Strategic Social Construction

Multinational corporations possess a preponderance of economic resources compared to other interest groups. Financial resources are crucial for maintaining lobbying efforts, building social networks, and more important, running campaigns. The cases at hand suggest that financial clout only mattered to a certain extent in the competition for political influence. There is no linear relationship between financial resources and influence. In the run-up to the Kyoto conference, the GCC was well organized and pooled substantial financial resources. While it could influence the Senate, it failed to block international agreement on mandatory emissions controls. At the time, the protrading coalition instead was only loosely organized and did not pool any financial resources. It relied on lobbyists from only a few core organizations such as BP and Environmental Defense. Yet it could shape policy in significant ways.

The financial and organizational resources of the protrading coalition, however, came to bear in setting up pilot trading schemes, as BP and Shell did later on in the process. BP's and Shell's global business organizations allowed them to run in-house trading schemes, thereby demonstrating the feasibility of a global trading regime. Hence, financial and organizational resources mattered significantly in generating the demonstration effect and expertise, which in turn were a source of discursive power for corporate climate leaders. Demonstration effects have proven to be crucial triggers of change in regulatory politics (Mattli and Woods 2009). In sum, the picture on the role of funding as a source of influence is mixed: to a certain extent financial resources are necessary to run advocacy campaigns, yet they are certainly not sufficient for advocacy success. Interesting questions flow from this: Under what conditions do financial resources translate into influence? And do financial resources have diminishing marginal returns?

Legitimacy: The Power of Compromise and Leadership

Legitimacy is the product of persuasion, social capital, or generally noncoercive forms of interaction. The case studies strongly confirm that transnational coalitions compete for legitimacy in the eyes of policymakers and the public. Legitimacy is the key political resource for access and influence. The empirical research has revealed three sources of the protrading coalition's superior degree of legitimacy: the provision of a compromise solution supported by business and NGOs, constructive leadership, and representativeness.

First, the protrading coalition gained strong legitimacy because it advocated a compromise formula that had the support of business and green groups. NGO-business cooperation was the appropriate strategy to exploit the policy crisis of climate politics. It can be argued that baptist-and-bootlegger-type collective action delivers interest consolidation through the negotiation of a compromise solution, which increases the legitimacy of the policy proposal. Some may assert that the term compromise is a euphemism for what actually was a co-optation of the environmental community by business. The case studies, though, suggest the opposite: environmental organizations made the conscious choice to promote market-based climate policy. Environmental Defense has been a champion for market-based environmental governance since the 1980s. Hence, legitimacy is granted intentionally to business and market-based policy solutions. Business was not co-opting environmental groups; the environmental community was split between those supporting market-based solutions and those preferring command-and-control policies. While classic baptist-and-bootlegger dynamics could be observed in the Kyoto and U.S. cases, the environmental community was not available as a coalition partner for business in the EU case. The European Commission's Directorate General for the Environment played the role of environmental advocate, thus granting legitimacy in the public's mind to the EU-wide trading scheme proposal. The fact that European environmental groups did not actively join protrading coalitions raises the general question as to whether environmental groups in the United States are more likely to enter NGO-business coalitions than environmental groups in Europe.

Second, policymakers listened to BP and other climate leaders because they demonstrated leadership by taking on emissions reduction targets, advocating a compromise policy solution, and setting up pilot trading experiments. This indicated goodwill and a constructive attitude toward policymakers. The credibility of climate leadership in the eyes of policymakers was enhanced by the fact that the ICCP in the United States and BP in the United Kingdom had a strong record of constructive engagement with government. With regard to the United States, the precedent of the Montreal Protocol played a significant role. A number of ICCP members were a constructive force in the Montreal Protocol's negotiations, earning them credibility with the U.S. administration. Leadership also generated expertise about the new policy instrument, which further enhanced the credentials of the protrading policy entrepreneurs.

Third, the legitimacy of the protrading coalition stemmed from the fact that it represented successful multinational corporations from critical economic sectors. It was important for all policy entrepreneurs to demonstrate that the advocacy coalitions represented large cross-sections of the economy. The support of economic powerhouses such as DuPont and General Electric signaled to policymakers that large parts of the mainstream of the business community were backing the policy. This was strikingly different from the early 1990s, when only so-called sunrise industries such as the renewable energy industry—small firms at the fringes of the economy—had called for climate regulation. It is noteworthy that legitimacy through representativeness prevailed over the logic of representativeness through large numbers. USCAP, for instance, rejected applications for membership, but carefully ensured that the coalition represented key economic sectors. Altogether the case studies show that business can acquire legitimacy in a number of ways, including through cooperation with NGOs, policy leadership, expertise, and cross-sectoral representativeness.

Coalition Strategies: State Allies and Multilevel Games

Next to the political environment and coalition resources, strategy is a critical variable in the equation of business influence. Power resources need to be skillfully employed in order to be able to exploit political opportunities. In fact, the case studies provide evidence that strategy is essential for advocacy success. The ability to mobilize state allies has especially proven to be indispensable for successful advocacy campaigns.

State Allies: The Sine qua non of Successful Advocacy

It is highly unlikely that protrading coalitions would have been successful in putting emissions trading on the international climate agenda had it not been for strong state allies. Indeed, in all three cases, a strong state ally pushed the instrument for its own reasons. In Kyoto, the Clinton administration promoted emissions trading. In the EU case, the European Commission was an ally for a market-based climate policy, while in the United States, states in the Northeast and California were trading advocates. The fact that state allies are crucial for advancing nonstate actor agendas suggests that the state remains the center of gravity in environmental politics (cf. Drezner 2007). Business organizes politically around state politics and allies with like-minded state actors. This observation on the centrality of the nation-state with regard to transnational business political action is echoed by studies on private governance. Many private governance arrangements emerge in the shadow of hierarchy to avoid mandatory regulation or are granted authority by the state (Falkner 2003; Mattli and Büthe 2005). This puts a question mark on notions of a governance gap that resulted from the state's retreat and that is now inhabited by business. It provides evidence for ideas of complex multilateralism that see the state system as the skeleton around which nonstate actors get involved (O'Brien 2003).

The fact that the policy preferences of the corporate protrading advocates converged with key state actors implies that advocacy is not only about changing policymakers' preferences but also about helping policymakers to legitimize a certain policy course. Business helped to mobilize broad-based support for market-based climate policy. With multinational corporations getting behind the proposal, policymakers had an easier case for justifying their policy. In this regard, it was important that coalitions helped to aggregate interests, as policymakers are looking for broad support for their policies. New cross-sectoral alliances such as the UK ETG and USCAP were a particularly effective form of interest aggregation as they spanned various economic sectors. It can be assumed that cross-sectoral interest aggregation is an effective strategy when a policy proposal affects a large share of an economy. Otherwise, the proposal might be easily dismissed as the agenda of a specific vested interest.

Furthermore, the emergence of alliances between segments of business as well as government demonstrates that both actor types are fragmented. With regard to states, splits occurred between branches of government (U.S. administration versus Congress) and across levels of government (European Commission versus member states; U.S. administration versus states). Hence, next to business and NGO conflicts, there is also internal state conflict. This internal fragmentation offers firms opportunities to ally with one government actor in the effort to persuade other government actors of their policy preference.

Multilevel Games: Affecting Global Change

The case studies support the proposition that successful transnational coalitions are capable of influencing political processes across political levels. Climate policy is not made in one place only but rather in many places, as the system of global climate governance is highly decentralized. As has been shown, the diffusion of emissions trading represents a series of political struggles, ranging from the subfederal to the national, regional, and international level. Without nonstate actor influence across these levels, it would have been unlikely that emissions trading would have diffused so rapidly. The protrading coalition could put market-based climate policy on the agenda of U.S., EU, and international climate politics. The antiregulatory coalition instead eventually lost the battle over mandatory climate policy at the international level and in the EU. Next to the diffusion of formal political authority, the diffuse nature of global agendas called for a multilevel strategy. Ideas for a global climate regime were promoted by a large number of political actors, including governments, international organizations, firms, and NGOs, to name a few. The lesson to draw from this is that the more dispersed formal and informal political authority is, the more important transnational collective action is for nonstate actors to achieve shifts in global agendas.

The case studies suggest that the ability to play multilevel games requires certain strategic capacities. The protrading coalition was characterized by a high degree of flexibility. While communication and information exchange occurred among a large array of players, the close coordination of political strategies among actors took place in more specific national or regional contexts. The coalition thus could capitalize on both the broad-based political support of a coalition of transnational scope and local political capital in a number of political systems. The largest challenge for multilevel strategies appeared to lie in a lack of transnational political capacities. For instance, the antiregulatory business forces in the EU failed to prevent mandatory climate policy because, among other reasons, they created an EU-wide coalition too late.

The case studies have also revealed a close link between the strategy to play multilevel games and venue shopping. Generally speaking, activists can target several political levels simultaneously or start with the most receptive venue to then move on to other ones. Tarrow (2005) has argued that transnational social movements typically emerge from a scaling up of local activism. Hence, activists first target their home governments before activism shifts to foreign countries and the international level. The case studies showed that the opposite also can be true, however. Protrading advocacy first emerged around the Kyoto Protocol, and later moved to the national or subnational level, to then scale up again to the regional or federal level.

Coalition building, to sum up, represents a strategic form of power through organization, as it allows coalitions to leverage sources of influence in three ways. First, coalitions pool both material and immaterial resources—notably financial resources and legitimacy. Second, coalitions allow for the broad aggregation of interests, which is key to mobilizing state allies for an agenda, as state actors look for broad support of their policy choice. Third, transnational coalitions hold the political infrastructure in place to be able to play political games at multiple levels ranging from the subnational to the international level. The level of influence of transnational business coalitions is an equation of political opportunities, power resources, and strategies, whereby this book has found that legitimacy, state allies, and the ability to play multilevel games are absolutely key. The finding that the mobilization of state allies—a crucial political strategy—represents a pivotal source of influence suggests that it is important to also consider sources of influence beyond the power resources owned by the firm, such as funding and lobbying networks. Strategy and political opportunities are also essential variables in business influence.

Beyond Climate: Transnational Business Coalitions in Other Issue Areas
While I do not attempt to test the analytic framework on coalition influence in other cases in this section, I will examine how prevalent and significant transnational business coalitions are in other areas of global environmental politics and beyond. Transnational advocacy networks and coalitions of civil society groups are well documented in the literature (Keck and Sikkink 1998; Khagram, Riker, and Sikkink 2001b; von Bulow 2010). This is not the case for transnational business coalitions. While there are a series of references to transnational business coalitions, in-depth case studies on individual coalitions are scarce. Studies exploring business collective action in environmental politics focus mostly on domestic industry coalitions (Vogel 1995; DeSombre 2000) or transnational rule-setting organizations (cf. Cutler, Haufler, and Porter 1999; Haufler 2001; Falkner 2003; cf. Djelic and Sahlin-Andersson 2006; Pattberg 2007). In the following, I discuss a small number of cases of transnational business coalitions in other issues areas to reflect on this book's findings in light of other cases.

In the realm of environmental politics, we could observe the emergence of transnational business coalitions in a number of cases, with ozone and biosafety being high-profile instances next to climate change (Newell and Levy 2006). In an effort to prevent further ozone regulations in the United States, the CFC industry created the Alliance for Responsible CFC Policy in 1980 (Haas 1992a; Falkner 2008).[2] Representing the interests of U.S. firms, the alliance first opposed the regulation of CFCs in the United States. Due to increasing scientific certainty about the ozone problem and shifting political dynamics, the coalition changed its strategy to support international caps on CFCs. DuPont led this political shift in 1986. Henceforth, DuPont and the Alliance for Responsible CFC Policy became the hub of a larger transatlantic coalition aiming to shape the international rules of ozone regulation. On the other side of the Atlantic, it was in particular the German chemical company Hoechst that pushed the new transnational industry agenda (Falkner 2008). The new proregulatory industry stance strengthened the position of the U.S. Department of State, which was in favor of an international agreement.

In biosafety politics, the Global Industry Coalition was the unified voice of biotechnology companies in Europe and the United States (Newell 2003). It also served as a forum for information exchange. Created in

1998, the coalition represented mostly seed companies, pharmaceutical companies, commodity traders, and food manufacturers (Reifschneider 2002). Association leaders from Canada, Europe, and the United States were at the forefront of the coalition, which claimed to represent more than 2,200 companies from over 130 countries (Clapp 2003). The transnational coalition allowed industry to speak with one voice on fundamental issues while governments—in this case, the EU and the United States—were divided. Over the course of the political process, the Global Industry Coalition did not succeed in presenting a unified industry position in biosafety politics. Conflict between different factions of industry—in particular, between pharmaceutical biotech firms and the agribiotech industry—undermined a global voice for industry (Falkner 2009).

Beyond environmental politics, the relevance of transnational business coalitions has been stressed with regard to a number of cases. First of all, broad permanent coalitions such as the ICC and the WBCSD are involved in a number of policy fields (Ronit and Schneider 2000). Beyond these broad coalitions, more issue-specific coalitions are, for instance, found in European market integration, transatlantic trade liberalization, and the international protection of intellectual property rights. A few industrialists in Europe in the early 1990s decided to revive Europe economically with an ambitious reform program, responding to a decade of declining growth. Led by Pehr Gyllenhammar, the CEO of Volvo, company leaders from across Europe set up the European Round Table of Industrialists (ERT) in 1983, with the assistance of the European Commission (Cowles 1995). A transnational CEO coalition, the ERT emerged as a leading advocate for an integrated European market. The ERT found a particularly strong ally for a unified market in the French government under François Mitterrand. Maria Cowles argues that the ERT was instrumental in setting the agenda for the single market program, although business advocacy was not the only factor leading to the passage of the Single European Act in 1992. The ERT represents one of the first instances of transnational business coalition formation in Europe.

In a similar vein, the Transatlantic Business Dialogue promoted the liberalization of trade between Europe and the United States (Coen and Grant 2001; Cowles 2001). The transatlantic business coalition was launched in a speech by U.S. Secretary of Commerce Ronald Brown in 1994; it was formally set up in a year later. In various working groups and committees, representatives from EU and US business worked toward transatlantic consensus on a series of issues of transatlantic business regulation and trade liberalization. While the coalition had been initiated, its agenda was soon driven by firms (Coen and Grant 2001). Similar to the ERT, the Transatlantic Business Dialogue relies on the involvement of CEOs.

In the case of intellectual property rights, the Intellectual Property Committee, a U.S.-based business group, was instrumental in crafting a transnational coalition of firms from the United States, Europe, and Japan that pushed for a multilateral agreement to protect intellectual property rights (Sell 2003; Sell and Prakash 2004). As Susan Sell and Aseem Prakash argue, the Agreement on Trade-Related Intellectual Property Rights of 1994 is largely a result of the transnational business campaign for the international protection of intellectual property rights. The authors suggest that a policy crisis and successful framing strategies were part of the success of the transnational business network. Technological progress had made the appropriation of intellectual property-based products and processes widespread, thereby creating a demand for policy. The network exploited this political window of opportunity by framing the issue as a matter of competitiveness, which resonated with broader discourses. Yet in another case related to intellectual property rights, the transnational business coalition lost to a transnational NGO coalition. The HIV/AIDS epidemic led NGOs to call for generic drugs to be made available at low cost in developing countries. NGOs maintained that the protection of intellectual property rights should not compromise measures to protect public health. The access campaign succeeded in bringing about policy changes to the Agreement on Trade-Related Intellectual Property Rights and at the domestic level in the United States mainly by successfully exploiting the HIV/AIDS crisis as a political opportunity.

The preceding cases of transnational business coalitions from a number of issue areas in global politics suggest a few provisional conclusions. Transnational business coalitions tend to originate in the United States, and from there, they scale up to become global. Moreover, their membership is mostly from the United States and Europe, thus making them both transnational and transatlantic coalitions. The cross-section of coalitions also demonstrates that coalitions can be either pro- or antiregulatory, or shift from one political goal to another over the course of their lifetime. Another recurring theme is the role of individuals and key organizations as political entrepreneurs that are instrumental in creating the coalition as well as devising its strategy. Finally, a number of cases have underscored the important role of policy crises and state allies as sources of influence. While this discussion provides anecdotal evidence of the relevance of transnational business coalitions in other areas of world politics and relevant sources of influence, the field warrants comparative studies to advance our systematic understanding of advocacy-related business collective action in the global realm. Next to the issue of influence, a critical question is under what conditions transnational coalitions emerge and how they can be sustained.

The Strengths and Weaknesses of Alternative Explanations

This book has argued for a neopluralist, coalition-centered perspective on the rise of emissions trading. The general theoretical assumption is that business groups compete and cooperate among themselves as well as with other political actors, such as NGOs and states, over influence in global politics. The analytic focus of the neopluralist approach on conflict and coalitions has proven to be helpful in detecting broad patterns of contention across political systems and political levels. It allowed for a high degree of sensitivity to actual historical patterns of conflict and alliance building across state and nonstate actors. The theoretical contribution of this book to the neopluralist program lies in conceptualizing transnational business coalitions and their sources of influence. It has been demonstrated that the actual influence of transnational coalitions flows from the interplay of complex opportunity structures, available power resources, and skillful strategy employed by political entrepreneurs. While I contend that protrading coalitions were a key component, I acknowledge that other factors played into the rise of emissions trading. In fact, neo-Gramscians, neoliberals, and constructivists contribute valuable insights that complement the narrative on the role of protrading business coalitions in the diffusion of emissions trading.

Capital-Centered Explanations

The neo-Gramscian reading of the rise of emissions trading suggests the emergence of a new historical bloc that is driven by global finance (Paterson and Newell 2008; Newell and Paterson 2010). In the face of pressure from states and the environmental movement, business reconfigured the historical bloc around emissions trading. Carbon trading represents a strategy to accommodate demand for environmental action while also creating business opportunities for financial services providers.

Although the overall neo-Gramscian argument on the alignment of material, organizational, and discursive structures in a historical bloc is intriguing, it is not fully supported by the empirical data. First, neo-Gramscians assume that hegemony is established through the alignment of the interests of powerful business groups and dominant state actors. This book has in fact frequently emphasized the role of state allies in advancing the protrading agenda. Emissions trading only emerged on the international agenda because the U.S. government insisted on it. Subsequently, however, the instrument diffused globally in the absence of the support of the single most powerful state in the international system. By the time that the EU and other countries were promoting a global carbon market, the United States had withdrawn from the Kyoto Protocol and there were no indications that it would implement mandatory climate policies in the near term. Levy and Egan (2003) acknowledge that the lack of support of the U.S. federal government for emissions trading, even though business was increasingly supporting it, remains puzzling. Nonetheless, it could be asserted that governments established an "internationalized state" through the agreement of the Kyoto Protocol, which substitutes for U.S. power.

Yet it is questionable whether an international regime could produce the degree of stability that neo-Gramscian notions of an internationalized state suggest (cf. Victor 2001). Randall Germain and Michael Kelly's general comment (1998, 27) on the pitfalls of neo-Gramscian international relations theory is particularly valid in climate politics: a "corresponding structure of concrete political authority" is missing at the international level. International agreements are not sufficient as a structure of political authority in order to become genuinely hegemonic. The lack of alignment between the U.S. government and influential parts of the business community calls into question whether it is appropriate to conceptualize the diffusion of emissions trading as the reconfiguration of a historical bloc. The political process has proven to be much more contingent than neo-Gramscian notions of hegemony would suggest.

Furthermore, neo-Gramscians claim that the structural power of global finance is driving the shift to market-based climate policy. This fits nicely with the notions of finance-driven economic globalization prominent in neo-Gramscian thinking. As the preceding case studies show, however, financial services providers were not among the first advocates of emissions trading. Big emitters from the oil industry and power sector instead were the first movers. They supported emissions trading as a strategy to manage regulatory risk. Large banks and investment houses only entered the political stage when the EU ETS was up and running. Only then did the idea of creating a new financial market mobilize firms that could benefit from the new regulation.

Neo-Gramscians might counter that financial services providers did not need to get involved in the political process for its structural power to have an effect on the outcome. The structural power of global finance works more subtly as policymakers factor it automatically into their decisions. If this was the case, the notion to promote the financial sector by establishing a global carbon market must have figured in governments' decisions to support emissions trading. With the notable exception of the United Kingdom, this was not the case. As in the minds of big emitters, most governments perceived emissions trading as the most cost-effective policy instrument to address climate change. Only over time did the idea of creating a huge financial market gain prominence. Corporate protrading advocacy thus was mainly part of a proregulatory risk management strategy as opposed to a proregulatory market-making strategy. In sum, the historical analysis of the three cases shows a number of political contingencies and patterns of agency that fit slightly uneasily with the neo-Gramscian narrative of the rise of emissions trading.

Notwithstanding this critique, there is also considerable overlap between a neo-Gramscian reading and a coalition-centered interpretation of the globalization of carbon trading. Both stress the role of alliances and coalitions in environmental policymaking in general and the global diffusion of market-based environmental policy in particular. The idea that business draws on the support of environmental groups and/or state allies is found in both approaches, albeit with different connotations. In a neo-Gramscian understanding, state allies emerge because governments realize what the interest of *capital-in-general* is. In a pluralist perspective, nonstate actors instead have to pursue state allies actively through, for instance, interest aggregation.

State-Centered Explanations

Neoliberal institutionalists would argue that the Kyoto Protocol resulted from intergovernmental negotiations between rationally acting states (Rowlands 2001). International institutions are considered to be the product of interest-based bargaining between parties (Barrett 2005). In this view, emissions trading was part of the Kyoto package as a deal between the United States and the EU. Agreement between the parties, moreover, was facilitated by the UNFCCC as an international institution. States subsequently complied with the international agreement, so the contention goes, by implementing it nationally. Does the empirical story of the diffusion of emissions trading mirror such a process? Neoliberal institutionalists are right in identifying states as key creators and carriers of international institutions. While nonstate actor advocacy played a significant role in all three cases, the rules for emissions trading schemes were ultimately decided and implemented by state actors at various levels. Carbon markets rely heavily on government intervention, as they are policy-driven markets. I have also demonstrated that the success of nonstate actor advocacy depends strongly on the mobilization of state allies, as governments remain the crucial decision makers.

Yet the contractarian strand of neoliberal institutionalism remains silent as to the formation of state interests. Why did the United States and the EU settle on international emissions trading? For years, negotiations between the two powers were caught in a stalemate. The United States opposed international emissions reduction mandates, while the EU was highly skeptical of market mechanisms as part of an international climate regime. In the run-up to Kyoto, however, the United States accepted targets and timetables, while the EU warmed to market mechanisms. The sudden shift in negotiating positions remains largely unexplained by the classic strand of neoliberal institutionalism. As has been shown, this shift is best interpreted as resulting from a change in the underlying political-economic coalitions. The antiregulatory coalition started to disintegrate and a new NGO-business coalition emerged, which helped to legitimize a U.S. foreign policy that supported mandatory yet market-based climate policy. While not commensurate with classic versions of neoliberal institutionalism, this analysis speaks to theoretical advances in neoliberal institutionalism that have led to the inclusion of domestic politics in the analysis of international cooperation (cf. Milner 1997; Simmons 1997).

As for the diffusion of emissions trading after the agreement of the protocol, the institutionalist notion of compliance has limited value. While the Kyoto Protocol allowed for the use of emissions trading, it did not prescribe its use in domestic climate policies. Nonetheless, the EU, once the most ardent opponent of market-based climate policy, pioneered a multilateral trading scheme. Since there was no obligation to comply, compliance could not have motivated the EU to develop the EU ETS. A number of other factors in fact played into the EU's decision to establish the EU ETS, including the protrading campaign of oil and power companies, but also the interests of the European Commission. Nonparties to the protocol also started to design emissions trading schemes—at odds with the notion that the global rise of carbon trading was a straightforward process of compliance to an international treaty. U.S. states in particular pushed ahead with designing regional emissions trading. This suggests that the post-Kyoto diffusion of emissions trading was a transnational process of norm diffusion, rather than a process of compliance to an international treaty. These two notions are not necessarily mutually exclusive. The Kyoto Protocol certainly created normative support for emissions trading, but international law was not the mechanism behind the actual spread of emissions trading schemes. Instead, nonstate actor advocacy was one of the main transmission belts of cap and trade in the post-Kyoto phase.

Ideational Explanations

From a constructivist perspective, emissions trading spread globally because market-based environmental policy represents a powerful idea that shows great fit with the underlying normative structures of the international system (Bernstein 2001). This is essentially a structural argument that "emphasizes the constitutive dimension of norms, wherein norms do not merely regulate behaviour, but define social identities and practices" (ibid., 182). Emissions trading is understood to be part of the norm set of "liberal environmentalism," which exhibits a strong degree of compatibility with the liberal economic order of the international system. In fact, one of the experts interviewed for this study explicitly said that emissions trading was an idea whose time had come. The idea itself resonated so strongly with the political environment that its institutionalization occurred quasi-automatically. It is, however, exactly the issue of timing that raises a question about the explanatory power of structural constructivist interpretations of the global rise of carbon trading.

The economic idea of carbon trading had emerged years before the instrument was seriously considered in climate politics (Stavins 2003). As detailed in chapter 3, the idea was discussed in U.S. academic circles for decades before it came to affect real-world policymaking. Moreover, its application to global climate change was suggested as early as the late 1980s. Not only the idea of emissions trading but also the underlying matching liberal normative order existed for decades. Hence, there was a strong match between an idea and the normative order without this idea institutionalizing in global politics. This suggests that the idea itself can barely have been the driving force, but that other factors played into the diffusion process. Indeed, the instrument only became an option for policy design when vested interests were strategically advocating emissions trading as a policy response to global warming. The United States, for instance, showed a great normative fit between market-based environmental policy and underlying norms of a liberal market order. But carbon trading did not gain ground until political forces shifted significantly in its favor in the mid-1990s and again around 2005.

Bernstein argues that liberal environmentalism is the global compromise position, which would suggest that emissions trading is a good match for countries around the world. The case studies send a note of caution regarding this assumption. What we find is rather an Anglo-American compromise related to liberal environmentalism. Emissions trading is very much a U.S. concept that resonates strongly with the institutional framework of liberal market economies in the United States and the United Kingdom, whereas the fit with coordinated market economies is suboptimal. The fact that emissions trading de facto globalizes does not imply that the instrument is compatible with institutional/normative settings across the globe. Rather, it tells a story about the power distribution in the process of policy diffusion.

These criticisms are not to say that the issue of a fit between policy ideas and underlying normative or institutional structures does not matter. As I argue, it represents a political opportunity structure for nonstate actors, but it does not explain the spread of an idea. Institutional/normative fit is a conducive, but not sufficient factor for the diffusion of ideas. The critique presented here is directed at structural mainstream variants of constructivism that consider ideas as explanatory variables. As maintained earlier, agency-oriented variants of constructivism converge with the neopluralist, coalition-centered assertion, as they build on rationalist assumptions (Goldstein and Keohane 1993; Price 1998). This book builds directly on this work, explicitly discussing the sources of influence of nonstate actor coalitions in struggles over meaning and processes of norm diffusion. Political-economic coalitions can be understood as the driver and carriers of ideas. Ideas such as carbon trading advance on the agenda because agents actively promote them using political resources and skills under given environmental conditions. To recall Peter Hall (1989), behind every economic idea there is an interest. Hence, ideas act as intervening variables between interests and agency-based forms of power, on the one hand, and political agendas and outcomes, on the other. This is said with a caveat in mind, though: once an idea has become institutionalized as a norm, structural notions of constructivism might well have considerable explanatory power. Yet the focus of this book is on the emergence and spread of a policy idea before it is strongly institutionalized in the normative order of the international system.

To conclude, the neo-Gramscian, institutionalist, and constructivist interpretations of the global rise of carbon trading leave a series of historical facts unexplained, but also contribute a number of complementary insights. Important additional insights relate to the role of states in setting the rules for carbon markets as well as discursive processes and structures. Some of these insights have been incorporated in the framework through the consideration of norms as opportunity structures and the focus on state allies as a source of influence for nonstate actors. Moreover, certain strands within these theoretical traditions show considerable overlap with neopluralism. These include the coalition-oriented analysis in neo-Gramscian approaches, institutionalist analyses of the role of domestic politics in state interest formation, and agency-oriented variants of constructivism. As argued earlier, the key explanatory contribution of the neopluralist, coalition-centered reading lies in its analytic sensitivity toward conflict and coalitions between nonstate and state actors across countries and political levels. It thus allows one to describe broad patterns of contention in global environmental politics. In analyzing conflict and coalitions, the neopluralist lens pays great attention to agency. Political entrepreneurs are critical in organizing either political conflict or cooperation.

The particular contribution of this book to debates on transnational neopluralism lies in advancing the conceptualization of transnational business coalitions. It lays out an original framework that considers global political context, power resources, and political strategy. Such an influence-oriented analytic framework allows us to understand why coalitions succeed or fail in a particular political context. At a broader level, the framework contributes to an influence-oriented perspective as opposed to a power-oriented one in the study of business in global environmental politics. The coalition-related contentions made in this book thus contribute to the neopluralist research program in the study of global environmental politics (Betsill and Corell 2008; Falkner 2008), and in international relations and international political economy more generally (Mattli and Woods 2009; Avant, Finnemore, and Sell 2010; Cerny 2010).

Business Influence and Market-Based Global Environmental Governance

What does the preceding story of the globalization of carbon trading tell us about the influence of business in the shift toward market-based environmental governance? In the following, I will describe the level of influence of business and discuss whether the rise of market-based climate policy is a case of regulatory capture. I will then offer a brief assessment of the environmental effectiveness of emissions trading.

Giving in to Social Demand, Shaping Regulatory Style

The influence of business extended to shaping the regulatory style of climate policy, but firms gave in to the social demand for mandatory climate legislation. The corporate campaign for emissions trading is partly a response to the mounting demand for mandatory emissions cuts in industrialized countries. While business had evidenced the political force to fend off mandatory climate regulation in the form of carbon-energy taxes in the early 1990s, such strong opposition to climate legislation proved increasingly to bear political and reputational costs. European companies in particular, such as BP and Shell, were conscious of alienating policymakers and consumers by continuing to oppose international emissions reduction mandates. By changing course to support a market-based international regime, business responded to the growing social demand for climate action. Firms shifted from an antiregulatory strategy to a proregulatory risk management strategy. Business support for a mandatory yet market-based regime enabled governments to agree on legally binding emissions reduction targets. In this situation, the protrading coalition was a vehicle for business to shape the style of mandatory climate legislation in its favor, ruling out carbon taxes and command-and-control policies. Hence, business had the influence to shape the regulatory style, but gave in to the growing social demand for mandatory emissions cutbacks.

Regarding the business preference for market-based policy, the empirical findings include a note of caution against generalizing that business is in favor of market-based environmental policy as such and everywhere. Market-based environmental policy is not per se business friendly. Though carbon taxes are a market-based instrument, industry opposed taxes as much as it opposed command-and-control standards, whereas emissions trading and voluntary agreements received business support. Within the toolbox of market-based environmental policy, emissions trading represents a special case. Due to the fact that it creates a new market and that permits can be allocated for free, it can mobilize the support of a great number of business actors. Furthermore, emissions trading was driven by firms from liberal market economies. These observations suggest that business support for market-based environmental policies varies considerably across countries, sectors, and individual instruments (cf. Jordan, Wurzel, and Zito 2003). While tradable permits display a good institutional fit in the United States and the United Kingdom, environmental taxes and voluntary agreements are widespread in continental Europe.

Normative Compromise Instead of Regulatory Capture

Business pioneered emissions trading and was pivotal in its diffusion. The prominent role of multinational corporations in the diffusion of emissions trading could be read as business single-handedly driving the shift toward market-based environmental governance, capturing governments and environmental organizations (cf. Newell 2008a). This is not the case. Governments and a few NGOs equally acted as policy entrepreneurs for emissions trading with their own interests in mind. These actors held their policy preferences independently from business influence. This predisposition of other key actors toward market-based policy created opportunities for the corporate protrading campaign.

Hence, business did not co-opt or capture governments and green groups into market-based climate policy. The market-based climate policy instead represents a normative Anglo-American compromise between factions of business, factions of the environmental community, and a few government actors (cf. Bernstein and Ivanova 2007).[3] The success of this liberal Anglo-American climate compromise raises questions about the role of the underlying power structures in the international system that favor Anglo-American models. While I suggest that the spread of emissions trading represents a global normative compromise between business and governments, there is strong evidence that the design process of cap-and-trade schemes has to some extent fallen prey to regulatory capture. The EU ETS is a case in point, where business lobbying led to the overallocation of emissions permits.

The Invisible Green Hand: Good or Bad?

The rise of market-based global environmental governance is highly contested and controversial: Is it for better or worse? This question divides governments, business, and green groups alike. On the one hand, supporters of market-based governance view private governance and environmental markets as mechanisms that increase the efficiency of environmental regulation and unleash the potential of the private sector in solving environmental problems (Holliday, Schmidheiny, and Watts 2002). On the other hand, critics claim that market-based governance empowers business vis-à-vis government, leading to a weakening of environmental regulation and a lack of democratic control. Market-based governance is said to have led to a crisis of global environmental governance by disembedding the global economy (Bernstein and Ivanova 2007; Park, Conca, and Finger 2008). The verdict is still out on the efficiency and effectiveness of private governance and environmental markets. While not discussing the effectiveness of market-based environmental policy in general, I offer a provisional assessment of the effectiveness of carbon trading.[4] The available data on the effectiveness of carbon trading are limited to the trial period of the EU ETS (cap and trade) and the CDM (baseline and credit).

The trial period of the EU ETS ran from 2005 to 2007. This first period has been criticized for mainly two flaws (Ellerman and Buchner 2007).

First, due to a lack of accurate emissions data, emissions permits to emitting entities were overallocated. The carbon price plummeted as a result. Second, electric utilities reaped significant windfall profits by passing along the costs of freely allocated allowances. The trial period of the EU ETS thus led to only small emissions reductions. The emissions of the sectors covered by the EU ETS flattened during the 2005–2007 period despite robust gross domestic product growth. Hence, the trial period was somewhat effective in terms of reducing CO_2 emissions.

Yet as A. Denny Ellerman and Paul Joskow (2008) argue, the trial period was not meant to lead to significant emissions reductions but rather to establish the trading scheme and provide lessons for reform. The EU ETS in fact delivered on these criteria. It established the market infrastructure and created a carbon price, which companies started to incorporate into their decision making. In December 2008, the EU passed a reform package, which aimed to make the system more effective. In particular, it introduced partial auctioning as the allocation method and granted the European Commission stronger authority in the allocation process. It also introduced a cap until 2020. The effectiveness of the scheme hinges critically on the ability of member states and the European Commission to manage the market. The performance of the second trading period so far provides cause for cautious optimism. Anecdotal evidence suggests that higher allowance prices in 2008 led to fuel switching in the power sector and improvements in the efficiency of power plants, which resulted in emissions reductions (Ellerman, Convery, and De Perthuis 2010).

The experience with the CDM is similar to that with the EU ETS: it underperformed regarding its environmental outcome, which is mostly due to design issues. Again, the imperfect performance is not surprising and does not question the instrument per se. The CDM pipeline has been operational since December 2003. The EU ETS has driven the demand for CDM credits.

Criticisms of the CDM relate mostly to its limited scope and the additionality of emissions reductions achieved through CDM projects (Harvard Project on International Climate Agreements 2009). The CDM's limited scope is due to a couple of reasons. First, the approval of CDM projects through the CDM executive board is a bureaucratic and expensive process. Every single project has to go through this approval process in order to receive credits. Second, the CDM excludes a number of mitigation activities such as the conservation of forests that could reduce GHG emissions, which limits the overall amount of emissions reductions achieved through the mechanism. With regard to additionality, it is an open question whether CDM projects are in fact additional to what would happen under a business-as-usual scenario. The CDM executive board has developed methodologies to determine a project's baseline emissions. Given the complexity of the issue and state of the methodologies, experts fear that a significant number of CDM projects would have been conducted in the absence of the CDM—that is, they are not additional (Schneider 2007).

Linked to this is the issue of what kinds of emissions-reducing projects are funded through the CDM. The CDM was meant to encourage investment into low-carbon energy infrastructure in developing countries (Wara 2007). Renewable energy projects, however, accounted for only 28 percent of the emissions reductions to be achieved through the CDM until 2012 (Wara 2009, 1779). The largest share of emissions reductions result from capturing and destroying industrial gases such as HFC-23, N_2O, and CH_4 emitted by landfills and confined-animal-feeding operations. Hence, in these cases the CDM credits did not spur investment in low-carbon energy infrastructure.

While the shortcomings of the CDM market are significant, a number of analysts suggest that institutional reform could greatly strengthen the mechanism (Victor and Cullenward 2007; Wara 2009). The CDM is important as it engages the fastest-growing economies such as China and India in global mitigation efforts.

In sum, both the EU ETS, the first multilateral cap-and-trade scheme, and the CDM, a global baseline-and-credit scheme, have produced only modest emissions reductions so far. Yet this has to be seen against the fact that the first few years of operation were trial periods, in which market infrastructure was established and lessons were learned. If institutional reform succeeds, there is a good chance that the carbon-trading schemes will score higher on environmental effectiveness (Grubb 2010). The governance challenges are not insurmountable (Victor and House 2004). Much of the performance hinges on how well government actors manage carbon markets. The results are mixed so far: the Copenhagen climate summit of 2009 (COP 15) fell short of providing a long-term, top-down framework for global carbon trading. Nevertheless, the reform of the EU ETS and the debate on reforming the CDM are cause for cautious optimism that governments learn relatively quickly when it comes to managing carbon markets.

Implications for the Study and Practice of Global Environmental Politics

This book's findings result in a number of conclusions on how to advance the study of business in global environmental politics. These relate to the role of business cooperation as a source of influence and the need to shift the analytic focus from power- to influence-centered studies of business. In addition, a number of practical implications for policymakers follow from the analytic findings. They revolve around the opportunities and risks emanating from the fact that business coalitions often act as the political-economic engines of global environmental politics.

Research Implications

This book speaks to the debate on the role of business in global politics in general and global environmental politics in particular. Only relatively recently did scholars start paying more attention to this field, moving slowly from description to theory making. This study of business influence in global climate politics suggests two major conclusions for future research on the role of business in global environmental politics.

First, the role of business in global environmental politics should be analyzed through the lens of both business conflict *and* business cooperation. While the former is a key factor diminishing business influence, the latter is a crucial source of business influence. Collective action in transnational coalitions, for instance, creates an organizational form of power that allows business to leverage power resources and collective political strategies. The research agenda on transnational business collective action could be pursued further in a number of directions. At the descriptive level, different types of collective action need to be differentiated. Transnational business coalitions are only one form of transnational collective action—though an increasingly prevalent one. In identifying types of transnational collective action, close attention needs to be paid to institutionalized and less institutionalized forms of cooperation, such as ad hoc coalitions. A crucial question in terms of explanation is under what conditions firms engage in transnational collective action (cf. Mahoney 2007). As much as it is important to understand the conditions under which business conflict emerges, it is also important to understand why there is sometimes business unity and cooperation (Mizruchi 1989). The framework outlined in chapter 2 discusses the economic effects of environmental regulation on firms, institutional pressure, and the role of coalition brokers as potential variables in coalition formation. These factors offer a preliminary avenue into this field, which needs to be taken further.

Second, the study of business in global environmental politics needs to shift from a power-oriented focus to an influence-oriented one. As I mentioned earlier, the concept of power refers to actor capabilities. A great deal of scholarly attention has been directed at the different dimensions of corporate power, such as instrumental, discursive, and structural power (Levy and Egan 1998; Fuchs 2005a; 2008; Falkner 2008). Yet power is just one variable in the equation of corporate influence. As the framework developed in chapter 2 reflects, influence is a function of power capabilities, strategic skills, and political opportunity structures. Power can, but must not, translate directly into influence in shaping political agendas and outcomes (Corell and Betsill 2008). In the end, though, influence is the more relevant category when we try to understand how nonstate actors affect the outcome of global environmental politics. A central question in this regard is under what environmental conditions the instrumental, discursive, and/or structural power of firms translates into influence.

In pursuing research in these directions, I argue that the study of business in global environmental politics will benefit from reaching across the theoretical aisle by borrowing from social movement theory. This strand of research has produced a considerable body of literature on transnational collective action along with the influence of nonstate actors and social movements (Zald 1996; Keck and Sikkink 1998; Tarrow 2005). Many of these insights can be transferred to the study of business in global environmental politics. In fact, a number of international relations scholars, including Karsten Ronit (2007) and John Ruggie (2004), have called for integrating the study of business and civil society to advance the theory making of collective action in the transnational sphere. At a minimum, the empirical reality itself is calling for such an approach, as firms and NGOs are increasingly engaging with each other in new forms of transnational collaboration, as shown in this book. There is, moreover, no theoretical reason to consider firms and NGOs as two distinct types of nonstate actors.

Much scholarship on transnational relations has advanced a bifurcated view in which interests motivate firms and norms motivate NGOs. On the basis of this, the two actors were studied separately. Yet interests and norms in reality play into the motivation of any actor (Sell and Prakash 2004). The immediate advantage of integrating research on firms and NGOs is theoretical cross-fertilization. Another advantage lies in the analytic aggregation of the study of nonstate actors in global politics, which will allow for the generalization of findings and theory making. Key questions cutting across a number of forms of transnational collective action can be addressed. These questions regard the emergence of transnational collective action, its organizational structure, the political strategies, and the sources of influence of transnational collective action. This strategy promises to move the debate beyond primarily descriptive accounts of transnational collective action toward theory. Erica Johnson and Aseem Prakash (2008) have suggested a similar way ahead for the study of NGOs.

Practical Implications

The story of global climate politics demonstrates that business support is crucial for the passage of major environmental policy. All attempts to enact climate regulation that had only the support of the environmental community, such as carbon taxes and command-and-control regulation, largely failed. Against this backdrop, the emergence of proregulatory coalitions appears to be an important factor for successful climate policymaking. While such coalitions present policymakers with opportunities, they also bear risks.

Unlike any other policy instrument, carbon trading has mobilized a wide range of actors from industry and the environmental community. Big emitters see it as an opportunity to minimize their compliance costs, financial services firms and technology providers stand to benefit from the market-creating effect of carbon trading, and green groups like the certainty that carbon trading offers regarding the quantity of emissions reductions. Carbon trading thus mobilized the political-economic forces that are essential for the successful enacting of climate policy and a shift to a low-carbon economy (Newell and Paterson 2010). The protrading coalition exhibits the classic features of baptist-and-bootlegger-style advocacy, where strange bedfellows join forces.

Some policy instruments lend themselves more to such dynamics than others. In the case of climate policy, a market-based policy instrument could mobilize industry constituencies. Yet it would be a false conclusion to assume that market-based environmental policy in general has such effects. Carbon taxes, for example, are also a market-based instrument, but found they little support among business. For business to support an environmental policy instrument, it is more important that the policy is the least costly option and/or that it has significant market-creating effects. Command-and-control policies might exhibit these properties in some cases, as, for instance, in the phaseout policies for CFCs. If this is a given, proregulatory alliances may emerge by themselves. Sometimes, though, government actors can intervene in the process of interest aggregation among firms and green groups to help the creation of alliances. Policymakers need to develop a keen eye for identifying potentials for such dynamics, and the skills to capitalize on them.

Still, the mobilization of business support for market-based climate policy is a double-edged sword. While firms may lend their support to the general policy principle, conflict with policymakers arises over the nitty-gritty details of the design of trading schemes. The devil lies literally in the detail. Critical design elements include the sectoral coverage of a trading scheme, the allocation method, and accounting methods, to name but a few. Big emitters especially have an interest in the free allocation of permits and lax rules to minimize compliance costs. What is more, the complexity of cap-and-trade schemes offers firms a number of entry points for lobbying and rent seeking. This has led to the real risk that actual cap-and-trade schemes fall prey to regulatory capture, rendering trading schemes potentially environmentally ineffective.

As I argued earlier, the last word on the effectiveness of global emissions trading has not yet been spoken. What is already clear, however, is that the success of carbon markets hinges critically on the ability of state actors to carefully manage them. It is the paradox of pollution markets that they harness market forces while creating a strong demand for skillful government intervention. This requires the ability to mobilize business forces, on the one hand, and partly insulate decision making on the design of carbon market from corporate influence, on the other. Policymakers are still in the early phase of the learning curve of managing pollution markets. It remains to be seen how successful they prove to be in embedding carbon markets in effective rules and regulations. Disembedded carbon markets would ultimately undermine the fragile compromise of market-based climate policy.

Notes

Chapter 1

1. For the purpose of this study, the terms emissions trading, carbon trading, market mechanisms, and flexible mechanisms are used interchangeably. Moreover, emissions trading is understood to include both cap-and-trade schemes and project-based mechanisms. This differs from the terminology of the Kyoto Protocol, which equates emissions trading with cap-and-trade schemes. The two different variants of emissions trading will be explained in detail in chapter 3.

2. Levy is a professor of management at the University of Massachusetts Boston College of Management. Kolk is a professor of sustainable management at the Amsterdam Business School.

3. The organization was formerly named the Environmental Defense Fund.

Chapter 2

1. This is the central argument of the "business conflict" model, which is a conceptual cornerstone of neopluralist approaches to business in global politics. Classic examples of this strand of research include the work of David Gibbs, Ronald Cox, Thomas Ferguson, Joel Rogers, Jeffrey Frieden, Helen Milner, David Abraham, and James Nolt.

2. The expression derives from the study of Prohibition, the ban of alcohol in the United States in the early twentieth century. At the time, Baptists were in favor of Prohibition because of moral reasons, while bootleggers supported it because they could sell alcohol to those who couldn't obtain it legally. This common ground led to the cooperation of unlike fellows and the Prohibition law.

Chapter 3

1. The question of trading versus taxation was vigorously debated in a forum on federal climate policy hosted by the Hamilton Project, an initiative of the Brookings Institution in Washington, DC.

2. Most of the emissions trading characteristics discussed here only relate to cap-and-trade rather than baseline-and-credits schemes.

3. Gilbert Metcalf (2007, 34) suggests that this might be "more urban legend than fact."

4. The White House staffers did not unanimously support emissions trading. John H. Sununu, the White House chief of staff, fiercely opposed the targets-and-timetables approach and emissions trading. These two views on emissions trading would persist throughout the history of climate politics.

5. Interviews are numbered for reasons of confidentiality.

6. If not indicated differently, all numbers in this section come from Point Carbon's report *Carbon 2008*.

7. The voluntary market for companies and consumers wanting to offset their GHG emissions is the second, but less significant driver of project-based trading mechanisms.

8. The founding members are Arizona, California, New Mexico, Oregon, and Washington State.

9. The participating states are Illinois, Iowa, Kansas, Michigan, Minnesota, and Wisconsin. Indiana, Ohio, and South Dakota have joined the initiative as observer states.

10. By August 2010, the International Carbon Action Partnership included the following members: the European Commission, France, Germany, Greece, Ireland, Italy, Netherlands, Portugal, Spain, the United Kingdom, Maine, Maryland, Massachusetts, New Jersey, New York, Arizona, British Columbia, California, Manitoba, New Mexico, Oregon, Quebec, Washington State, New Zealand, Australia, Norway, and the Tokyo Metropolitan Government. Observers were Japan, the Republic of Korea, and Ukraine.

Chapter 4

1. The Climate Council worked particularly on behalf of U.S.-based oil firms. It coalesced closely with the Organization of Petroleum Exporting Countries and was organized by Don Pearlman, a former undersecretary in the Reagan administration who worked for the Washington law firm Patton, Bloggs, and Pow.

2. The political influence of the oil industry in the United States transcends political parties and ideologies, contends Ran Goel (2004).

3. After Rio, the ratification of the convention progressed quickly, with the United States being among the first countries to ratify the convention. By December 1993, fifty countries had ratified the UNFCCC, thus fulfilling the criteria for the treaty's entry into force.

4. Founded in 1989, CAN acts as an umbrella for environmental organizations in international climate politics. In 2003, the network had more than 280 members and operated globally.

5. Already before the split in the oil industry, there was business conflict between the mainstream of business and so-called sunrise industries. The latter are

industries such as the renewable sector that would directly profit from climate regulation. Yet due to the minimal political clout of the new industries, this form of business conflict was largely irrelevant for policy outcomes.

6. It is notable that DuPont was involved in both sides of the business conflict. As an ICCP member, it supported the Kyoto framework. Through its subsidiary Conoco, it also sat in the meetings of the API, a crucial group fighting emissions controls.

7. BusinessEurope was formerly called UNICE.

8. The Kyoto Protocol received considerable support from EU industry, as a poll of the members of the United Kingdom's Confederation of British Industry in November 1997 revealed. Fully 83 percent of the member companies expressed support for the EU's 15 percent target, and 62 percent believed that Europe should pursue the goal unilaterally if the other industrialized nations refused to adopt it (Leggett 2001).

9. Shell's proactive strategy included in particular the launch of Shell International Renewables as a fifth core business in 1997 (Skjaerseth and Skodvin 2003).

10. Several U.S. firms from different industries followed the first movers in Europe. In late 1999, Ford Motor Company left the GCC, and was soon followed by DaimlerChrysler, General Motors, and Texaco.

11. There is a significant difference between the ICCP's statement in July 1996 and a press release from March 1996. In the earlier statement, the ICCP (1996a) did not yet mention emissions trading but instead called for a process that "preserves maximum national flexibility in achieving the commitments."

12. JUSCANNZ is an informal group of states—and an acronym for them—that includes Japan, the United States, Switzerland, Canada, Australia, Norway, and New Zealand. In international negotiations, other countries sometimes align with this group.

13. This was the first shift in the official position, although the shift in the EU's position began already in 1996. In February 1996, EU environment commissioner Ritt Bjerregard acknowledged that joint implementation might be an acceptable policy instrument of international cooperation.

14. Formerly called the Business Council for a Sustainable Energy Future.

15. Other critical concerns for the United States included the level of emissions reductions, the participation of developing countries, and the issue of sinks and the EU bubble, which let the EU member states meet their target collectively (interview 45).

16. Gore took this step against the recommendations of his political advisers (interviews 39 and 45). Since he was likely to be the Democratic candidate in the next election, they did not want him to be associated with the Kyoto Protocol. Reportedly, the vice president said: "I worked on this issue of climate change all my life. There is nothing I care more about than this. And if I can make a contribution to getting this done, I am willing to take the risk—whatever the political consequence is. I am going to Kyoto" (interviews 45).

17. Members included Australia, Canada, Iceland, Japan, New Zealand, Norway, Ukraine, and the United States. It evolved from the JUSSCANNZ group.

18. Between resigning from government service and becoming the executive director of the Pew Center, Claussen worked for the lobbying firm Alcalde and Fay, which had organized the ICCP (interview 33). This gives anecdotal insight into the close personal ties between the brokers of the protrading NGO-business coalition.

19. The GHG protocol has become the dominant standard for GHG accounting. Among other schemes, the EU ETS, the UK ETS, the CCX, and the California Climate Action Registry have adopted the standard.

20. The WBCSD's members were split over emissions trading, which is why IETA was set up as a separate organization. Key advocates among WBCSD's members were BP, Shell, DuPont, TransAlta, Tokyo Electric, and others (interview 51).

21. UNCTAD and Sandor, who helped set up the U.S. trading scheme for SO_2 emissions, tried to set up a voluntary trading scheme prior to Kyoto regardless of whether Kyoto would endorse trading (Boulton and Clark 1997).

22. Marcu had been a personal assistant to Strong, chair of Ontario Hydro and UNCED's secretary general, which reflects the close personal ties between the business elite at the core of the coalition that was promoting market-based environmental policy (interview 2).

23. The Environmental Markets Association is a U.S.-based trade association that was founded in 1996 in the context of SO_2 trading in the United States.

24. The dismantling of the GCC reflected the decline of the antiregulatory coalition. Some also perceived it as a sign of its success, however, that it had prevented the establishment of domestic targets for the United States. The Byrd-Hagel resolution and Bush's withdrawal from Kyoto are both responses to the GCC's lobbying efforts. In that sense, it is also a case of "mission accomplished," as an anonymous source in the oil industry stated (quoted in Levy and Egan 2003, 825).

25. Companies only started pooling financial resources in the IETA to strengthen their political clout after the network had been built.

Chapter 5

1. In other European countries, domestic trading schemes were being considered at the time. Both Sweden and Norway had constituted parliamentary commissions to assess the option of a domestic emissions trading scheme.

2. The UK ETS was a voluntary scheme, combining elements from both cap-and-trade and baseline-and-credit schemes.

3. The target converged broadly with the targets that countries had taken on under the Kyoto Protocol. While less ambitious than the EU's 15 percent reduction target below 1990s levels, it was more aggressive than the stabilization target of the United States (Victor and House 2006).

4. It is worth mentioning that the UK Treasury, which has a general predisposition toward taxes, commissioned the Marshall report.

5. Many of the early movers had considerable experience with trading. While BP and Shell always had trading desks, the power companies had gained trading expe-

rience in the recent past, since the United Kingdom's electricity market was liberalized a decade earlier than electricity markets in the rest of Europe (interview 41).

6. BP also financed an administrative assistant, and the UK government seconded John Craven from the Department of Trade and Industry to the UK ETG. He later became head of its secretariat.

7. Beyond those immediate motives, some companies had "parallel motives." As one company representative said, the UK ETG was also largely about supporting a policy project of the UK government to maintain good relations with the government. This mattered in particular to companies operating in highly regulated markets, where the design of regulations has a direct effect on a company's bottom line (interview 47).

8. Already in 1997, John Prescott, the UK deputy prime minister, urged the City of London to learn from Chicago in order to profit from the emerging lucrative market in GHG emissions (Boulton 1997). Chicago had become the hub of the SO_2 emissions market.

9. The UK ETG also set up a legal liaison group for lawyers. The attempt to establish a financial services liaison group was unsuccessful due to the limited interest of financial services providers in the carbon market at the time (interview 48).

10. While BusinessEurope officially supported emissions trading, it never emerged as a primary advocate. The association remained essentially ambivalent about the approach, as different industry factions were blocking each other in policy formulation.

11. German power companies had negotiated a voluntary agreement with the German government. In addition, the position of the German power sector was heavily influenced by coal-based utilities such as RWE. These utilities feared that a price on carbon would put them out of business.

12. This was a collaborative project between Eurelectric, the International Energy Agency, and ParisBourse. While GETS 1 included only power generators, the following GETS also covered other economic sectors.

13. Interestingly, emissions trading was met with the same arguments in the United States when it was first introduced in the Acid Rain Program. This suggests that there are not fundamental differences between European and U.S. NGOs as to their positioning on market-based instruments. U.S. NGOs had only been exposed to the instruments at an earlier point in time.

14. In general, in 1998 and 1999, the European debate on emissions trading relied on the import of expertise from the United States. Next to NGOs, officials from the EPA came to Europe to talk about lessons from SO_2 trading. Moreover, the Center for Clean Air Policy, a Washington-based think tank, was a key adviser in designing the EU ETS, making available the expertise it had gained from the sulfur-trading scheme. Soon the debate became tailored to the European context and the case of CO_2, however, which limited the opportunities for knowledge transfer (Zapfel and Vainio 2002).

15. This is unlike the United States, where the U.S. Department of State holds the official lead on climate change policy.

16. A communication is the weakest form of a proposal by the European Commission to the Council of the European Union and the European Parliament.

17. Bill Kyte (Powergen) and Margaret Mogford (British Gas), two prominent architects of the UK trading scheme, alternately represented the UK ETG at the stakeholder group's meetings.

18. The electricity sector was represented by the Vereinigung der Elektrizitätswerke, which was a member association of Eurelectric. The chemical sector had considerable influence in the BDI, thus dominating the climate policy position of German industry as a whole. The BDI is one of the most powerful members of BusinessEurope.

19. Germany was offered an opt-out clause that allowed for the exemption of some installations in the beginning of the trading scheme. Furthermore, the country was allowed to allocate emissions rights for free and to whomever it favored. Finally, companies were granted the right to pool their emissions permits.

20. The group represented industry, government, and nongovernmental constituencies under the auspices of the environment ministry.

21. The directive's final text was published on October 13, 2003.

22. The second EU ETS trading period (2008–2012) coincides with the first commitment period under the Kyoto Protocol.

23. Next to the stringency of the regime, sectoral coverage was a key issue. The debate revolved around whether the aviation industry should be included in the EU ETS (Kirwin 2003).

24. ECIS was set up in October 2006. Initially it had eighteen members comprising thirteen investment banking groups, three Kyoto project developers, an exchange, and an international law firm. In May 2007, the international equivalent to ECIS was set up under the name International Climate Change Finance and Services Association.

25. Interestingly, the Vereinigung der Elektrizitätswerke, the association of the German power companies, once was fiercely opposed to emissions trading, but the experience with the EU ETS has led to a U-turn.

26. The EU member states here are France, Germany, Greece, Ireland, Italy, Netherlands, Portugal, Spain, and the United Kingdom. The members of the U.S. Regional Greenhouse Gas Initiative here are Maine, Maryland, Massachusetts, New Jersey, and New York. The Western Climate Initiative participants here are Arizona, British Columbia, California, Manitoba, New Mexico, Oregon, and Washington State.

Chapter 6

1. Barbour lobbied for several other clients, among them the industry coalition Electric Reliability Coordinating Council. The group included, for instance, Duke Energy.

2. The link between campaign contributions and influence over climate policy remains inconclusive. For instance, Kenneth Lay, the CEO of Enron, an energy

company, organized a fund-raising event for Bush that raised US$21.3 million for his campaign (Lisowski 2002). Enron was a strong supporter of the Kyoto Protocol, although that obviously did not affect Bush's decision.

3. U.S. members of the protrading network were the exception. Most of them gathered in the BELC.

4. The full program title is Voluntary Innovative Sector Initiatives: Opportunities Now.

5. This company was later renamed Environmental Financial Products LLC.

6. DuPont was particularly active in designing the CCX (interview 29).

7. Beyond its original U.S. operations, CCX became an influential player in the global carbon market when it launched the European Climate Exchange in 2005. The European Climate Exchange emerged as the leading exchange in the EU ETS.

8. The Pew Center and the WRI were involved in the process as advisers, whereas Environmental Defense acted as an advocacy group like NRDC (interview 49).

9. Many other business groups and individual companies were involved (see also http://www.rggi.org/design/history/stakeholder_meetings). The groups mentioned, however, were the key business groups.

10. With this step Schwarzenegger intensified California's earlier efforts in climate change policy. In 2002, California governor Gray Davis had already mandated CO_2 caps for motor vehicles, causing a legal battle between the state's government, industry, and the federal government.

11. Pavley had already authored AB 1493, which regulates tailpipe emissions and was passed in 2002.

12. Environmental Entrepreneurs is a national group of environmentally oriented business professionals that works with the NRDC in organizing business support for environmental policy. Bob Epstein, who has been credited with a key role in passing AB 32, was the group's leader at the time. The Silicon Valley Leadership Group represents more than two hundred Silicon Valley companies, mainly from high-tech sectors.

13. The companies represented included BP, DuPont, Virgin Group, Google, Edison International, Swiss Re, Interface, AEP, British Sky Broadcasting, Pratt Industries, PG&E, Timberland, Goldman Sachs, JP Morgan Chase, and others (BP 2006).

14. Cinergy merged with Duke into Duke Energy in 2005–2006. Jim Rogers, the former CEO of Cinergy, became the CEO of Duke Energy.

15. By October 2007, the coalition was comprised of twenty-seven companies from a number of industries and six environmental groups.

16. While USCAP received administrative support from consultancies that also facilitated discussions, it has no full-time staff.

17. Beyond a few advocates, a carbon tax never received strong political support in the United States. The few economists calling for a carbon tax are referred to as the Pigou group (interview 43).

18. Other groups that once were at the heart of the antiregulatory camp are moving toward a more proactive stance to be able to shape the rules. Among these are the chemical and coal industries.

19. It is noteworthy that in the House of Representatives, two proposals for a carbon tax were introduced: Stark (H.R. 2069) and Larson (H.R. 3416).

20. Reportedly, the launch of USCAP drew Warner's attention to the climate change issue (interview 44).

21. Among the reasons for the defeat were the high prices for energy, the economic crisis, and the 2008 general election.

Chapter 7

1. Such transnational learning and the role of business in policy diffusion potentially offers clues into the cross-fertilization of varieties of capitalism, which can be observed as coordinated market economies are increasingly adopting elements from liberal market economies. Furthermore, next to cross-fertilization via business, the case study points to the role of public actors as transmission belts between varieties of capitalism, or more generally speaking, between different institutional settings. Unlike the majority of European governments, the European Commission was inclined to further market-based environmental governance. Second only to British firms, the commission essentially acted as a gatekeeper for emissions trading to Europe.

2. It was later renamed the Alliance for Responsible Atmospheric Policy.

3. The term compromise might evoke positive connotations, although a normative judgment is not intended. The term is used strictly in the descriptive sense that parties from all three camps were in favor of market-based policy, and promoted it as a third way next to the more radical policy goals of the antiregulatory business coalition and the mainstream of the environmental movement.

4. Environmental effectiveness is only one criterion among others for assessing the value of carbon trading.

References

ACBE. 1998. *Climate Change: A Strategic Issue for Business*. London: Advisory Committee on Business and the Environment.

Adler, Emanuel. 1997. Seizing the Middle Ground: Constructivism in World Politics. *European Journal of International Relations* 3:319–363.

Alcock, Frank. 2008. Conflicts and Coalitions within and across the ENGO Community. *Global Environmental Politics* 8 (4): 66–91.Andrews, Kenneth T., Michael Ganz, Matthew Baggetta, Hahrie Han, and Chaeyoon Lim. 2010. Leadership, Membership, and Voice: Civic Associations That Work. *American Journal of Sociology* 115 (4): 1192–1242.

Arimura, Toshi H., Dallas Burtraw, Alan J. Krupnick, and Karen L. Palmer. 2007. U.S. Climate Policy Developments. Discussion paper. Washington, DC: Resources for the Future.

Arts, Bas. 1998. *The Political Influence of Global NGOs: Case Studies on the Climate and Biodiversity Conventions*. Utrecht: International Books.

Asselt, Harro van, and Frank Biermann. 2007. European Emissions Trading and the International Competitiveness of Energy-Intensive Industries: A Legal and Political Evaluation of Possible Supporting Measures. *Energy Policy* 35 (1): 497–506.

Avant, Deborah D., Martha Finnemore, and Susan K. Sell, eds. 2010. *Who Governs the Globe?* Cambridge: Cambridge University Press.

Bachram, Heidi. 2004. Climate Fraud and Carbon Colonialism: The New Trade in Greenhouse Gases. *Capitalism, Nature, Socialism* 15 (4): 1–16.

Bang, Guri, Andreas Tjernshaugen, and Steinar Andresen. 2005. Future U.S. Climate Policy: International Re-engagement? *International Studies Perspectives* 6 (2): 285–303.

Barker, Colin, Alan Johnson, and Michael Lavalette. 2001. Leadership Matters: An Introduction. In *Leadership and Social Movements*, ed. Colin Barker, Alan Johnson, and Michael Lavalette, 1–23. Manchester: Manchester University Press.

Barnett, Michael, and Raymond Duvall. 2005. Power in International Relations. *International Organization* 59 (1):39–75.

Baron, Richard, and Michel Colombier. 2005. Emissions Trading under the Kyoto Protocol: How Far from the Ideal? In *Climate Change and Carbon Markets: A*

Handbook of Emissions Reduction Mechanisms, ed. Farhana Yamin, 153–165. London: Earthscan.

Barrett, Scott. 2005. *Environment and Statecraft: The Strategy of Environmental Treaty-Making*. Oxford: Oxford University Press.

Barringer, Felicity. 2007. A Coalition for Firm Limit on Emissions. *New York Times*, January 19, 1.

Baumgartner, Frank R., and Bryan D. Jones. 1991. Agenda Dynamics and Policy Subsystems. *Journal of Politics* 53 (4): 1044–1074.

Baumgartner, Frank R., and Bryan D. Jones. 1993. *Agendas and Instability in American Politics*. Chicago: University of Chicago Press.

Baumgartner, Frank R., and Beth L. Leeth. 1998. *Basic Interests: The Importance of Groups in Politics and in Political Science*. Princeton, NJ: Princeton University Press.

Bendell, Jem. 2000. *Terms for Endearment: Business, NGOs, and Sustainable Development*. Sheffield, UK: Greenleaf Publishing.

Bernstein, Steven. 2001. *The Compromise of Liberal Environmentalism*. New York: Columbia University Press.

Bernstein, Steven, and Benjamin Cashore. 2000. Globalization, Four Paths of Internationalization, and Domestic Policy Change. *Canadian Journal of Political Science* 33 (March): 67–99.

Bernstein, Steven, and Maria Ivanova. 2007. Institutional Fragmentation and Normative Compromise in Global Environmental Governance: What Prospects for Re-Embedding. In *Global Liberalism and Political Order: Toward a New Compromise*, ed. Steven Bernstein and Louis W. Pauly, 161–185. Albany: State University of New York Press.

Berry, Jeffrey M. 1989. *The Interest Group Society*. New York: HarperCollins.

Betsill, Michele M. 2008a. Environmental NGOs and the Kyoto Protocol Negotiations: 1995 to 1997. In *NGO Diplomacy*, ed. Michele M. Betsill and Elisabeth Corell, 43–66. Cambridge, MA: MIT Press.

Betsill, Michele M. 2008b. Reflections on the Analytical Framework and NGO Diplomacy. In *NGO Diplomacy: The Influence of Nongovernmental Organizations in International Environmental Organizations*, ed. Michele M. Betsill and Elisabeth Corell, 177–206. Cambridge, MA: MIT Press.

Betsill, Michele M., and Elisabeth Corell. 2001. NGO Influence in International Environmental Negotiations: A Framework for Analysis. *Global Environmental Politics* 1 (4): 65–85.

Betsill, Michele M., and Elisabeth Corell, eds. 2008. *NGO Diplomacy: The Influence of Nongovernmental Organizations in International Environmental Negotiations*. Cambridge, MA: MIT Press.

Betsill, Michele M., and Matthew J. Hoffmann. 2008. The Evolution of Emissions Trading Systems for Greenhouse Gases. Paper presented at the Annual Convention of the International Studies Association, San Francisco, March 26–29.

Biello, David. 2003. A Revolutionary Season. *Environmental Finance*, July–August 2003, 18–20.

Bodansky, Daniel. 2001. The History of the Global Climate Change Regime. In *International Relations and Global Climate Change*, ed. Urs Luterbacher and Detlef F. Sprinz, 23–40. Cambridge, MA: MIT Press.

Bode, Sven. 2005. Emissions Trading Schemes in Europe: Linking the EU Emissions Trading Scheme with National Programs. In *Emissions Trading for Climate Policy*. Bernd Hansjürgens, 199–221. Cambridge: Cambridge University Press.

Boulton, Leyla. 1997. Clinton Tries to Calm Storm over Climate: Reaching a Deal with US Industry on Greenhouse Gas Emissions Is Crucial to Kyoto Talks. *Financial Times*, October 6, 5.

Boulton, Leyla, and Bruce Clark. 1997. Emissions Traders Set Out Their Stall. *Financial Times*, October 21, 4.

Boulton, Leyla, and Bethan Hutton. 1997. Kyoto Climate Change Talks Agree Treaty at Last Minute. *Financial Times* (North American ed.), December 11, 1.

BP. 2006. BP Hosts Blair and Schwarzenegger Climate Change Meeting. Press release. London, July 31.

Bradsher, Keith, and Andrew C. Revkin. 2001. A Pre-emptive Strike on Global Warming: Many Companies Cut Gas Emissions to Head Off Tougher Regulations. *New York Times*, May 15, 1.

Broder, John, and Jad Mouawad. 2009. Energy Firms Find No Unity on Climate Bill. *New York Times*, October 18, 1.

Brown, Kevin and Andrew Taylor. 1999. Ministers under Pressure to Scrap Energy Tax. *Financial Times*, March 28, 11.

Browne, John. 1997. *Addressing Global Climate Change: Speech on May 19, 1997*. Palo Alto, CA: Stanford University Press.

Buntin, John. 1999. *Cleaning up the "Big Dirties": The Problem of Acid Rain.* Cambridge, MA: Kennedy School of Government.

Burstein, Paul. 1991. Policy Domain: Organization, Culture, and Policy Outcomes. *Annual Review of Sociology* 17:327–350.

Busch, Per-Olof, and Helge Jörgens. 2005. The International Sources of Policy Convergence: Explaining the Spread of Environmental Policy Innovations. *Journal of European Public Policy* 12 (5): 860–884.

Business Challenge. 1997. A Business Climate Challenge: America Needs to Get Serious about Global Warming. *Wall Street Journal*, December 1, A17.

Business Community Calls for Government Action on Climate Change Now. 2007. Press release. Available at <http://www.e5.org/downloads/G8BusinessCallForActionOnClimateChange1.pdf> (accessed October 4, 2010).

California Assembly. 2006. Global Warming Solutions Act. AB 32. Available at http://www.leginfo.ca.gov/pub/05-06/bill/asm/ab_0001-0050/ab_32_bill_20060927_chaptered.pdf (accessed January 13, 2011).

California First with Global Warming Control Legislation. 2006. *Environment News Service*, August 31. Available at http://www.ens-newswire.com/ens/aug2006/2006-08-31-04.html (accessed January 13, 2011).

Campbell, John L. 2005. Where Do We Stand? Common Mechanisms in Organizations and Social Movements Research. In *Social Movements and Organization Theory*, ed. Gerald F. Davis, Doug McAdam, W. Richard Scott, and Mayer N. Zald, 41–68. Cambridge: Cambridge University Press.

Capoor, Karan, and Philippe Ambrosi. 2007. *The State and Trend of the Carbon Market 2007*. Washington, DC: World Bank.

Carraro, Carlo, and Christian Egenhofer. 2003. Introduction to *Firms, Governments, and Climate Policy: Incentive-Based Policies for Long-Term Climate Change*, 1–16. Cheltenham, UK: Edward Elgar.

Cerny, Philip G. 1990. *The Changing Architecture of Politics: Structure, Agency, and the Future of the State*. London: Sage.

Cerny, Philip G. 2003. The Uneven Pluralization of World Politics. In *Globalization in the Twenty-First Century: Convergence or Divergence?* ed. Axel Hulsemeyer, 153–175. Basingstoke, UK: Palgrave Macmillan.

Cerny, Philip G. 2010. *Rethinking World Politics: A Theory of Transnational Neopluralism*. New York: Oxford University Press.

Chameides, William, and Michael Oppenheimer. 2007. Carbon Trading Over Taxes. *Science* 315 (March 23): 1670.

Changing the Climate of Opinion. 2000. *Economist*, December 8. Available at http://www.economist.com/node/12052171 (accessed January 13, 2011).

Checkel, Jeffrey T. 1993. Ideas, Institutions, and the Gorbachev Foreign Policy Revolution. *World Politics* 45 (2):271–300.

Checkel, Jeffrey T. 1998. The Constructivist Turn in International Relations Theory. *World Politics* 50 (2): 324–348.

Checkel, Jeffrey T. 2006. Tracing Causal Mechanisms. *International Studies Review* 8 (2): 362–370.

Christiansen, Atle C., and Jorgen Wettestad. 2003. The EU as a Frontrunner on Greenhouse Gas Emissions Trading: How Did It Happen and Will the EU Succeed? *Climate Policy* 3 (1): 3–18.

Cinergy. 2004. *Air Issues Report to Stakeholders*. Cincinnati: Cinergy Corporation.

Clapp, Jennifer. 2003. Transnational Corporate Interests and Global Environmental Governance: Negotiating Rules for Agricultural Biotechnology and Chemicals. *Environmental Politics* 12 (4): 1–23.

Clapp, Jennifer, and Doris Fuchs. 2009a. Agrifood Corporations, Global Governance, and Sustainability: A Framework for Analysis. In *Corporate Power in Global Agrifood Governance*, ed. Jennifer Clapp and Doris Fuchs, 1–25. Cambridge, MA: MIT Press.

Clapp, Jennifer, and Doris Fuchs, eds. 2009b. *Corporate Power in Global Agrifood Governance*. Cambridge, MA: MIT Press.

Clark, Ian. 2003. Legitimacy in a Global Order. *Review of International Studies* 29 (1) :75–95.

CLGCC. 2005. Letter to the Prime Minister. Available at<http://www.cpsl.cam.ac.uk/pdf/Letter%20to%20PM%202005%20Final.pdf> (accessed October 4, 2010).CLGCC. 2006. Business Leaders Support Tougher Stance on Climate Change; Offer to Work with Prime Minster to Make UK World Leader. Press release. Available at<http://www.cpsl.cam.ac.uk/leaders_groups/clgcc/uk_clg/press_releases/letter_uk_pm_june_06.aspx> (accessed October 4, 2010).

CLGCC. 2007. The Bali Communiqué on Climate Change. Available at <http://www.princeofwales.gov.uk/content/documents/Bali%20Communique.pdf> (accessed October 4, 2010).

Coase, Ronald H. 1960. The Problem of Social Cost. *Journal of Law and Economics* 3 (October): 1–44.

Coen, David. 1998. The European Business Interest and the Nation-state: Large-Firm Lobbying in the European Union and Member State. *Journal of Public Policy* 18 (1): 75–100.

Coen, David. 2005. Environmental and Business Lobbying Alliances in Europe: Learning from Washington? In *The Business of Global Environmental Governance*, ed. David L. Levy and Peter J. Newell, 197–220. Cambridge, MA: MIT Press.

Coen, David. 2007. Empirical and Theoretical Studies in EU Lobbying. *European Journal of Public Policy* 14 (3): 333–345.

Coen, David, and Wyn Grant. 2001. Corporate Political Strategy and Global Public Policy: A Case Study of the Transatlantic Business Dialogue. *European Business Journal* 13 (1): 37–44.

Cogan, Douglas G. 2006. Corporate Governance and Climate Change: Making the Connection. A Ceres report. Available at http://www.ceres.org//Document.Doc?id=90 (accessed January 13, 2011).

Cogan, Douglas G. 2008. *Corporate Governance and Climate Change: The Banking Sector. Boston: CERES report.*

Collier, Robert. 2007. "Cap and Trade" Gaining Favor: Congress Taking Up Business-Friendly Proposals to Reduce Global Warming. *San Francisco Chronicle*, 1.

Collingwood, Vivien. 2006. Non-Governmental Organisations, Power, and Legitimacy in International Society. *Review of International Studies* 32 (3): 439–454.

Commission on Collision Course over Greenhouse Gas Trading. 2001. *ENDS Report* 321 (October): 48–49.

Convery, Frank J. 2009. Reflections: The Emerging Literature on Emissions Trading in Europe. *Journal of Environmental Economics and Policy* 3 (1):121–137.

Cook, Steven D. 2001. U.S. Companies Weight Pros, Cons of Bush Decision to Pull Out of Kyoto Pact. *International Environment Reporter* 24 (17): 698–699.

Cook, Steven D. 2007. Bingaman Draft Bill Would Limit Emissions of Greenhouse Gases with Trading Program. *Environment Reporter* 38 (2): 69.

Cooper, Graham. 1999a. UK Unveils Hybrid GHG Trading Scheme. *Environmental Finance*, October, 4–5.

Cooper, Graham. 1999b. US Shuns World Bank's Carbon Fund. *Environmental Finance*, October, 14–16.

Corell, Elisabeth, and Michele M. Betsill. 2008. Analytical Framework: Assessing the Influence of NGO Diplomats. In *NGO Diplomacy: The Influence of Nongovernmental Organizations in International Environmental Negotiations*, ed. Michele M. Betsill and Elisabeth Corell, 19–42. Cambridge, MA: MIT Press.

Corporate Europe Observatory. 2000. *Greenhouse Market Mania: UN Climate Talks Corrupted by Corporate Pseudo-Solutions*. Amsterdam: Corporate Europe Observatory.

Cowles, Maria G. 1995. Setting the Agenda for a New Europe: The ERT and the EC 1992. *Journal of Common Market Studies* 33 (4): 501–526.

Cowles, Maria G. 2001. The Transatlantic Business Dialogue and Domestic Business-Government Relations. In *Transforming Europe: Europeanization and Domestic Change,* ed. M. G. Cowles, J. A. Caporaso and T. Risse-Kappen, 159–179. Ithaca: Cornell University Press.

Cox, Robert W., and Timothy J. Sinclair. 1996. *Approaches to World Order*. Cambridge: Cambridge University Press.

Cox, Ronald W., ed. 1996a. *Business and the State in International Relations*. Boulder, CO: Westview Press.

Cox, Ronald W. 1996b. Introduction: Bringing Business Back. In *Business and the State in International Relations*, ed. Ronald W. Cox, 1–7. Boulder, CO: Westview Press.

Crocker, Thomas D. 1966. The Structuring of Atmospheric Pollution Control Systems. In *The Economics of Air Pollution*, ed. Harold Wolozin, 61–68. New York: W. W. Norton and Company.

Cutler, A. Claire, Virginia Haufler, and Tony Porter, eds. 1999. *Private Authority and International Affairs*. Albany: State University of New York Press.

Dahl, Robert. 1957. The Concept of Power. *Behavioral Science* 2 (3): 201–215.

Dales, John H. 1968. *Pollution, Property, and Prices*. Toronto: University of Toronto Press.

Debate on EU Emissions Trading Steps into Overdrive. 2003. ENDS Report 341 (June), 18–22.

Department for the Environment, Transport, and the Regions. 2000. A Greenhouse Gas Emissions Trading Scheme for the United Kingdom. Consultation paper. November.

Depledge, Joanna. 2005. *The Organization of Global Negotiations: Constructing the Climate Change Regime*. London: Earthscan.

DeSombre, Elizabeth R. 2000. *Domestic Sources of International Environmental Policy: Industry, Environmentalists, and U.S. Power*. Cambridge, MA: MIT Press.

DiMaggio, Paul J., and Walter D. Powell. 1991. The Iron Cage Revisited: Institutional Isomorphism and Collective Rationality in Organizational Fields. In *The*

New Institutionalism in Organizational Analysis, ed. Walter D. Powell and Paul J. DiMaggio, 41–62. Chicago: University of Chicago Press.

Djelic, Marie-Laure, and Kerstin Sahlin-Andersson, eds. 2006. *Transnational Governance: Institutional Dynamics of Regulation*. Cambridge: Cambridge University Press.

Dombey, Daniel and Vanessa Houlder. 2002. Greenhouse Gas Trading Looks Set to Balloon: Agreement on EU-Wide System Is Likely to Be Reached Next Week. *Financial Times*, June 12, 14.

Doremus, Paul N., William Keller, Louis W. Pauley, and Simon Reich. 1998. *The Myth of the Global Corporation*. Princeton, NJ: Princeton University Press.

Drezner, Daniel W. 2007. *All Politics Is Global: Explaining International Regulatory Regimes*. Princeton, NJ: Princeton University Press.

Dunn, Seth. 2005. Down to Business on Climate Change: An Overview of Corporate Strategies. In *The Business of Climate Change: Corporate Responses to Kyoto*, ed. Kathryn Begg, Frans van der Woerd, and David L. Levy, 31–46. Sheffield, UK: Greenleaf Publishing.

ECIS. 2006. Announcement of Second Phase Allocation Plans. Press release. Available at <http://www.climatechangecapital.com/news-and-events/press-releases/announcement-of-second-phase-national-allocation-plans.aspx> (accessed October 4, 2010).

EEI. 2007. EEI Global Climate Change Principles. Available at <http://www.eei.org/ourissues/TheEnvironment/Climate/Documents/070208_climate_principles.pdf> (accessed October 4, 2010).

Egenhofer, Christian. 2007. The Making of the EU Emissions Trading Scheme: Status, Prospects, and Implications for Business. *European Management Journal* 25 (6): 453–463.

Eilperin, Juliet, and Steven Mufson. 2008. Climate Bill Underlines Obstacles to Capping Greenhouse Gases. *Washington Post*, January 6. Available at http://www.washingtonpost.com/wp-dyn/content/article/2008/05/31/AR2008053102471.html (accessed January 13, 2011).

Ellerman, A. Denny, and Barbara K. Buchner. 2007. The European Union Emissions Trading Scheme: Origins, Allocation, and Early Results. *Review of Economics and Policy* 1 (A): 66–87.

Ellerman, A. Denny, Frank J. Convery, and Christian De Perthuis, eds. 2010. *Pricing Carbon: The European Emissions Trading Scheme*. Cambridge: Cambridge University Press.

Ellerman, A. Denny, and Paul L. Joskow. 2008. *The EU Emissions Trading System in Perspective*. Washington, DC: Pew Center on Global Climate Change.

Ellerman, A. Denny, Paul L. Joskow, and David Harrison Jr. 2003. *Emissions Trading in the U.S.: Experience, Lessons, and Considerations for Greenhouse Gases*. Washington, DC: Pew Center on Global Climate Change.

Elster, Jon. 1998. A Plea for Mechanisms. In *Social Mechanisms: An Analytical Approach to Social Theory*, ed. Peter Hedström and Richard Swedberg, 45–73. Cambridge: Cambridge University Press.

Emissions Trading Breakthrough Heralds "Birth of Carbon Economy." 2003. ENDS Report 342 (July).

Environmental Defense. 2000. Global Corporations and Environmental Defense Partner to Reduce Greenhouse Gas Emissions. Press release. January 10. Available at <http://www.edf.org/article.cfm?ContentID=503> (accessed October 4, 2010).

Environmental Defense. 2007. US Climate Action Partnership Doubles, More CEOs Call for Cap on Carbon. Press release. May 8. Available at <http://www.edf.org/pressrelease.cfm?ContentID=6348> (accessed October 4, 2010).

Epstein, Edwin. 1969. *The Corporation in American Politics*. Englewood Cliffs, NJ: Prentice Hall.

EurActiv. 2007. Lobbyists Scramble for Attention on Eve of Summit. March 7. Available at<http://www.euractiv.com/en/pa/lobbyists-scramble-attention-eve-summit/article-162291> (accessed October 4, 2010).

Eurelectric. 2000. GETS 2 Report. Brussels: Eurelectric.

Eurelectric. 2003. Eurelectric Preliminary Comments on Linking the Kyoto Protocol's Project Mechanisms with the EU Emissions Trading Scheme. Available at <http://www2.eurelectric.org/ DocShareNoFrame/Docs/1/EOIDKGHADLHKHODIANMBFIIGODVOHT6G9T18ADBNB6W3/Eurelectric/docs/DLS/LetterJI-CDM3006031-2003-431-0008-2-.pdf> (accessed October 4, 2010).

European Commission. 1998. Climate Change: Towards an EU Post-Kyoto Strategy. Communication (98) 353.

European Commission. 2000a. Green Paper on Greenhouse Gas Emissions Trading within the European Union. Communication (87) final.

European Commission. 2000b. *Mandate: Working Group 1 of the European Climate Change Programme*. Brussels: Flexible Mechanisms.

European Commission. 2001a. *Final Report: ECCP Working Group 1 "Flexible Mechanisms."* Brussels: European Commission.

European Commission. 2001b. *Green Paper on Greenhouse Gas Emissions Trading within the European Union: Summary of Comments*. Brussels: European Commission.

Fahrenthold, David. 2009. Utility Leaving U.S. Chamber over Stance on Climate Change. *Washington Post*, September 22. Available at http://www.washingtonpost.com/wp-dyn/content/article/2009/09/22/AR2009092203258.html (accessed January 13, 2011).

Falkner, Robert. 2000. *The Role of Firms in Global Environmental Politics: The Case of Ozone Layer Protection*. Oxford: Oxford University Press.

Falkner, Robert. 2003. Private Environmental Governance and International Relations: Exploring the Links. *Global Environmental Politics* 3 (2): 72–87.

Falkner, Robert. 2005. The Business of Ozone Layer Protection. In *The Business of Global Environmental Governance*, ed. David L. Levy and Peter J. Newell, 105–196. Cambridge, MA: MIT Press.

Falkner, Robert. 2008. *Business Power and Conflict in International Environmental Politics*. Basingstoke, UK: Palgrave Macmillan.

Falkner, Robert. 2009. The Troubled Birth of the Biotech Century. In *Corporate Power in Global Agrifood Governance*, ed. Jennifer Clapp and Doris Fuchs, 225–251. Cambridge, MA: MIT Press.

Fay, Kevin. 2005. *U.S. Perspectives on Emissions Trading and Climate Policy: European and American Business Perspectives on Emissions Trading and Climate Policy*. Washington, DC: Mistra Foundation and Resources for the Future.

Finnemore, Martha. 1996. Norms, Culture, and World Politics: Insights from Sociology's Institutionalism. *International Organization* 50 (2): 325–347.

Finnemore, Martha, and Kathryn Sikkink. 1998. International Norm Dynamics and Political Change. *International Organization* 52 (4): 887–917.

Finnemore, Martha, and Kathryn Sikkink. 2001. Taking Stock: The Constructivist Research Program in International Relations and Comparative Politics. *Annual Review of Political Science* 4:391–416.

Fligstein, Neil. 1996. Markets as Politics: A Political-Cultural Approach to Market Institutions. *American Sociological Review* 61 (4): 656–673.

Fligstein, Neil. 1997. Social Skill and Institutional Theory. *American Behavioral Scientist* 40 (4): 397–405.

Fligstein, Neil. 2002. *The Architecture of Markets: An Economic Sociology of Twenty-First Century Capitalist Societies*. Princeton, NJ: Princeton University Press.

Fuchs, Doris. 2005a. Commanding Heights? The Strength and Fragility of Business Power in Global Politics. *Millennium: Journal of International Studies* 33 (3): 771–801.

Fuchs, Doris. 2005b. *Understanding Business Power in Global Governance*. Baden-Baden: Nomos.

Fuchs, Doris. 2008. *Business Power in Global Governance*. Boulder, CO: Lynne Rienner Publishers.

Gallon, Gary. 1999. GHG Emissions Trading Group Launched in London. *Gallon Environment Letter* 3 (18).

Ganz, Marshall. 2000. Resources and Resourcefulness: Strategic Capacity in the Unionization of California Agriculture, 1959–1966. *American Journal of Sociology* 105 (4): 1003–1062.

Ganz, Marshall. 2009. *Why David Sometimes Wins: Leadership, Organization, and Strategy in the California Farm Worker Movement*. Oxford: Oxford University Press.

Gardiner, David. 2006. *Best Practices in Climate Change Risk Analysis for the Electric Power Sector*. Boston: Ceres.

Garner, Lynn. 2008. Manufacturers to Stress Energy Efficiency over Emissions Trading in Climate Debate. *Environment Reporter* 39 (22): 1054.

GCC. 1997. Vice President Gore Abandons U.S. Workers and Ignores Calls for Achievable Climate Policy from U.S. Business and Industry at Kyoto Climate Conference. Press release. Washington, DC: Global Climate Coalition.

Gelbspan, Ross. 2004. *Boiling Point: How Politicians, Big Oil and Coal, Journalists, and Activists Are Fueling the Climate Crisis—and What We Can Do to Avert Disaster*. New York: Basic Books.

George, Alexander L., and Andrew Bennett. 2004. *Case Study and Theory Development in the Social Sciences*. Cambridge, MA: MIT Press.

Germain, Randall D., and Michael Kenny. 1998. Engaging Gramsci: International Relations Theory and the New Gramscians. *Review of International Studies* 24 (1): 3–21.

German Industry Slams EU Emissions Trading Plan. 2001. Reuters, August 28.

Getz, Kathleen A. 1997. Research in Corporate Political Action. *Business and Society* 36 (1): 32–72.

Giddens, Anthony. 2009. *The Politics of Climate Change*. Cambridge: Polity Press.

Gill, Stephen R., ed. 1993. *Gramsci, Historical Materialism, and International Relations*. Cambridge: Cambridge University Press.

Gill, Stephen R., and David Law. 1989. Global Hegemony and the Structural Power of Capital. *International Studies Quarterly* 33 (4): 475–499.

Goel, Ran. 2004. A Bargain Born of a Paradox: The Oil Industry's Role in American Domestic and Foreign Policy. *New Political Economy* 9 (4): 467–492.

Goldstein, Judith, and Robert O. Keohane, eds. 1993. *Ideas and Foreign Policy: Beliefs, Institutions, and Political Change*. Ithaca, NY: Cornell University Press.

Goodfellow, Melanie. 2005. In Search of Certainty: Environmental Finance. Available at <http://www.environmental-finance.com> (accessed March 29, 2009).

Gorman, Hugh S., and Barry D. Solomon. 2002. The Origins and Practice of Emissions Trading. *Journal of Political History* 14 (3): 293–320.

Government of Canada. 2008. Turning the Corner: Taking Action to Fight Climate Change. Available at <http://www.ec.gc.ca/doc/virage-corner/2008-03/brochure_eng.html> (accessed December 9, 2008).

Granovetter, Mark. 1985. Economic Action and Social Structure: The Problem of Embeddedness. *American Journal of Sociology* 91 (3): 481–510.

Green, Jessica F. 2008. The Regime Complex for Emissions Trading: The Role of Private Authority. Paper presented at the Annual Convention of the International Studies Association, San Francisco, March 25 -29.

The Greening of Wall Street. 2008. *Economist*, March 13. Available at http://www.economist.com/node/10855003 (accessed January 13, 2011).

Greenwood, Justin. 2007. *Interest Representation in the European Union*. Basingstoke, UK: Palgrave Macmillan.

Grice, Martin 1999. BP Amoco Goes Global with Emissions Trading. *Environmental Finance*, October, 5, 5.

Griffin, Murray. 2006. Leading Australian Companies Join Forces to Push for Legislation on Carbon Emissions. *Environment Reporter* 37 (15): 790.

Grubb, Michael. 1989. *The Greenhouse Effect: Negotiating Targets*. London: Royal Institute of International Affairs.

Grubb, Michael, Tim Laing, Thomas Counsell, and Catherine Willan. 2010. Global Carbon Mechanisms: Lessons and Implications. Climatic Change. January 16. Available at http://www.springerlink.com/content/ev02737p83172m28/ (accessed January 13, 2011).

Grubb, Michael, Christiaan Vrolijk, and Duncan Brack. 1999. *The Kyoto Protocol: A Guide and Assessment*. London: Earthscan.

Haas, Peter M. 1992a. Banning Chlorofluorocarbons: Epistemic Community Efforts to Protect Stratospheric Ozone. *International Organization* 46 (1): 187–224.

Haas, Peter M. 1992b. Introduction: Epistemic Communities and International Policy Coordination. *International Organization* 46 (1): 1–35.

Haas, Peter M., and Emanuel Adler. 1992. Conclusion: Epistemic Communities, World Order, and the Creation of a Reflective Research Program. *International Organization* 46 (1): 368–390.

Haas, Peter M., Robert O. Keohane, and Marc A. Levy. 1993. *Institutions for the Earth: Sources of Effective International Environmental Protection*. Cambridge, MA: MIT Press.

Haggard, Stephan. 1990. *Pathways from the Periphery*. Ithaca, NY: Cornell University Press.

Hall, Peter A. 1989. *The Political Power of Economic Ideas*. Princeton, NJ: Princeton University Press.

Hall, Peter A., and David Soskice, eds. 2004. *Varieties of Capitalism: The Institutional Foundations of Comparative Advantage*. Oxford: Oxford University Press.

Hall, Rodney Bruce, and Thomas J. Biersteker. 2002. *The Emergence of Private Authority in Global Governance*. Cambridge: Cambridge University Press.

Hansjürgens, Bernd. 2005. Introduction to *Emissions Trading for Climate Policy*, 1–16. Cambridge: Cambridge University Press.

Harvard Project on International Climate Agreements. 2009. Options for Reforming the Clean Development Mechanism. *Issue Brief* 2009-1. Cambridge, MA: Harvard Project on International Climate Agreements.

Harvey, Fiona, and John Aglionby. 2007. Business Lobby Demands Clear Emissions Goals. *Financial Times* (London), November 12. Available at http://www.ft.com/cms/s/0/77b0861a-a743-11dc-a25a-0000779fd2ac.html (accessed January 13, 2011).

Haufler, Virginia. 2001. *A Public Role for the Private Sector: Industry Self-Regulation in a Global Economy*. Washington, DC: Carnegie Endowment for International Peace.

Heclo, Hugh. 1978. Issue Networks and the Executive Establishment. In *The New American Political System*, ed. Anthony King, 87–124. Washington, DC: American Enterprise Institute.

Heinz, John P., Edward O. Laumann, Richard L. Nelson, and Robert H. Salisbury. 1993. *The Hollow Core: Private Interests in National Policymaking*. Cambridge, MA: Harvard University Press.

Helm, Burt. 2005. Cinergy Answers Burning Questions. December 12. Available at <http://www.businessweek.com/magazine/content/05_50/b3963413.htm> (accessed October 4, 2010).

Hertsgaard, Mark. 2006. CA Leads on Climate. *Nation*, September 14: 5–6.

Hillman, Amy H., Gerald D. Keim, and Douglas Schuler. 2004. Corporate Political Activity: A Review and Research Agenda. *Journal of Management* 30 (6): 837–857.

Hoffman, Andrew J. 2001. *From Heresy to Dogma: An Institutional History of Corporate Environmentalism*. Stanford, CA: Stanford University Press.

Hoffman, Andrew J. 2006. *Getting Ahead of the Curve: Corporate Strategies That Address Climate Change*. Arlington, VA: Pew Center on Global Climate Change.

Holliday, Charles O., Jr., Stephan Schmidheiny, and Philip Watts, eds. 2002. *Walking the Talk: The Business Case for Sustainable Development*. Sheffield, UK: Greenleaf Publishing.

Hooghe, Liesbet, and Gary Marks. 2003. Unraveling the Central State, But How? Types of Multi-Level Governance. *American Political Science Review* 97 (2): 233–243.

Hopgood, Stephen. 2003. Looking Beyond the "K-Word": Embedded Multilateralism in American Foreign Environmental Policy. In *US Hegemony and International Organizations: The United States and Multilateral Institutions*, ed. Rosemary Foot, S. Neil MacFarlane, and Michael Mastanduno, 139–166. Oxford: Oxford University Press.

Houlder, Vanessa. 1999a. A Bull Market in Hot Air: Trading of Rights to Produce Greenhouse Gases Is Being Held Back by Political Doubts. *Financial Times*, November 4, 22.

Houlder, Vanessa. 1999b. Oiling the Wheels of Change: BP Amoco Wants to Stand Apart from Its Peers but Recognises the Dangers of Hubris. *Financial Times* (North American ed.), April 15, 4.

Houlder, Vanessa. 1999c. The World Warms to Emissions Market. *Financial Times* (North American ed.), February, 4.

Houlder, Vanessa. 2001. Business Welcomes Emissions Trading Scheme. *Financial Times*, August 15, 3.

Houlder, Vanessa. 2003. EU Paves Way for Emissions Trading. *Financial Times*, June 26, 10.

Hovi, Jon, Tora Skodvin, and Steinar Andresen. 2003. The Persistence of the Kyoto Protocol: Why Other Annex I Countries Move on without the United States. *Global Environmental Politics* 3 (4): 1–23.

How Green Is Browne? 1999. *Economist*, April 17. Available at http://www.economist.com/node/199636 (accessed January 13, 2011).

Hula, Kevin W. 1999. *Lobbying Together: Interest Group Coalitions in Legislative Politics*. Washington, DC: Georgetown University Press.

ICCP. 1996a. *International Industry Coalitions Urges Governments and Industry to Join to Create Effective Global Climate Change Process: Press Release*. Washington, DC: International Climate Change Partnership.

ICCP. 1996b. *ICCP Supports U.S. Call for Long-Term Focus on Climate Change Issue: Press Release*. Washington, DC: International Climate Change Partnership.

ICCP. 2006. *Oppose unless Amended Position on AB32: ICCP Letter to California Legislature, 15/08/06*. Washington, DC: International Climate Change Partnership.

ICCP. 2007. ICCP Welcomes Climate Policy Dialogue. Press release. Available at <http://www.iccp.net/docs/uscap.pdf> (accessed October 4, 2010).

Immelt, Jeff. 2005. *Global Environmental Challenges: Speech at George Washington University*, September 05. Washington, DC: George Washington University.

Innovest. 2005. *Carbon Disclosure Project 2005*. London: Carbon Disclosure Project.

International Institute for Sustainable Development. 1997. Report of the Third Conference of the Parties to the United Nations Framework Convention on Climate Change. *Earth Negotiations Bulletin* 12 (76).

The Invisible Green Hand. 2002. *Economist*, June 7. Available at http://www.economist.com/node/1200205 (accessed January 13, 2011).

Jacob, Klaus, Marian Beise, Jürgen Blazejczak, Dietmar Edler, Rüdiger Haum, Martin Jänicke, Thomas Löw, Ulrich Petschow, and Klaus Rennings. 2005. *Lead Markets for Environmental Innovations*. Heidelberg: Physica-Verlag.

Jehl, Douglas, and Andrew C. Revkin. 2001. Bush, in Reversal, Won't Seek Cut in Emissions of Carbon Dioxide. *New York Times*, March 14. Available at http://www.nytimes.com/2001/03/14/us/bush-in-reversal-won-t-seek-cut-in-emissions-of-carbon-dioxide.html (accessed January 13, 2011).

Jessop, Bob. 1990. *State Theory: Putting Capitalist States in their Place*. Cambridge, UK: Polity Press.

Johnson, Erica, and Aseem Prakash. 2008. NGO Research Program: A Collective Action Perspective. Paper presented at the Annual Convention of the International Studies Association, San Francisco, March 25–29.

Jones, Bryan D. 2001. *Politics and the Architecture of Choice: Bounded Rationality and Governance*. Chicago: University of Chicago Press.

Jones, Charles A., and David L. Levy. 2007. North American Business Strategies towards Climate Change. *European Management Journal* 25 (6): 428–440.

Jones, Stephen C., and Paul R. McIntyre. 2007. Filling the Vacuum: State and Regional Climate Change Initiatives. *Environment Reporter* 38 (30): 1640.

Jordan, Andrew, Rüdiger K. W. Wurzel, and Anthony R. Zito 2003. "New" Environmental Policy Instruments: An Evolution or Revolution in Environmental Policy? *Environmental Politics* 12 (1): 201–224.

Joskow, Paul L., and Richard Schmalensee. 1998. The Political Economy of Market-Based Environmental Policy: The US Acid Rain Program. *Journal of Law and Economics* 41 (1): 37–83.

Keck, Margaret E., and Kathryn Sikkink, eds. 1998. *Activists beyond Borders: Advocacy Networks in International Politics*. Ithaca, NY: Cornell University Press.

Keohane, Nathaniel O., Richard L. Revesz, and Robert N. Stavins. 1998. The Choice of Regulatory Instruments in Environmental Policy. *Harvard Environmental Law Review* 22 (2): 313–368.

Keohane, Robert O. 1993. Institutional Theory and the Realist Challenge after the Cold War. In *Neorealism and Neoliberalism*, ed. David A. Baldwin, 269–300. New York: Columbia University Press.

Keohane, Robert O., and Joseph S. Nye. 1972. *Transnational Relations and World Politics*. Cambridge, MA: Harvard University Press.

Keynes, John Maynard. 1936. *The General Theory of Employment, Interest, and Money*. London: Macmillan.

Khagram, Sanjeev, James V. Riker, and Kathryn Sikkink. 2001a. From Santiago to Seattle: Transnational Advocacy Groups. In *Restructuring World Politics: Transnational Social Movements, Networks, and Norms*, ed. Sanjeev Khagram, James V. Riker, and Kathryn Sikkink, 3–23. Minneapolis: University of Minnesota Press.

Khagram, Sanjeev, James V. Riker, and Kathryn Sikkink, eds. 2001b. *Restructuring World Politics: Transnational Social Movements, Networks, and Norms*. Minneapolis: University of Minnesota Press.

King, Gary, Robert O. Keohane, Sidney Verba. 1994. *Designing Social Inquiry: Scientific Inference in Qualitative Research*. Princeton, NJ: Princeton University Press.

Kingdon, John. 1995. *Agenda, Alternatives, and Public Policies*. New York: Longman.

Kirwin, Joe. 2001. European Industry Continues to Oppose Proposed Mechanism for Emissions Trading. *International Environment Reporter* 24 (21): 861.

Kirwin, Joe. 2003. EU Environment Ministers Give Approval to COs Trading Scheme under Kyoto Accord. *International Environment Reporter* 26 (1): 9–10.

Kitschelt, Herbert. 1986. Political Opportunity Structures and Political Protest: Anti-Nuclear Movements in Four Democracies. *British Journal of Political Science* 16 (1): 57–85.

Knox-Hayes, Janelle. 2009. The Developing Carbon Financial Service Industry: Expertise, Adaptation, and Complementarity in London and New York. *Journal of Economic Geography* 9 (6): 749–777.

Kossoy, Alexandre, and Philippe Ambrosi. 2010. *State and Trends of the Carbon Market 2010*. Washington, DC: World Bank.

Kranhold, Kathryn. 2007. GE's Environment Pupsh Hits Business Realities. *Wall Street Journal*, September 14, A1, A10.

Krauss, Clifford, and Kate Galbraith. 2009. Climate Bill Splits Exelon and U.S. Chamber. *New York Times*, September 29. Available at http://www.nytimes.com/2009/09/29/business/energy-environment/29chamber.html (accessed January 13, 2011).

Krugman, Paul. 2002. Ersatz Climate Policy. *New York Times*, February 15. Available at http://www.nytimes.com/2002/02/15/opinion/ersatz-climate-policy.html (accessed January 13, 2011).

Krupp, Fred. 1986. New Environmentalism Factors in Economic Needs. *Wall Street Journal*, November 20, 34.

Kyoto Aftermath: Big Battles Still Ahead as U.S. Holds Treaty Key. 1997. *Oil and Gas Journal*, December 12, 17.

Labatt, Sonia, and Rodney R. White. 2007. *Carbon Finance: The Financial Implications of Climate Change*. Hoboken, NJ: John Wiley and Sons.

Langrock, Thomas. 2006. The Role of Stakeholder-Driven Corporate Governance: The Example of BP's Climate Change Strategy. In *Emissions Trading and Business*, ed. Ralf Antes, Bernd. Hansjürgens, and Peter Letmathe, 241–255. Heidelberg: Physica Verlag.

Lawrence, Thomas B. 1999. Institutional Strategy. *Journal of Management* 25 (2): 161–188.

Layzer, Judith A. 2007. Deep Freeze: How Business Has Shaped the Global Warming Debate in Congress. In *Business and Environmental Policy: Corporate Interests in the American Political System*, ed. Michael E. Kraft and Sheldon Kamieniecki, 93–125. Cambridge, MA: MIT Press.

A Lean, Clean Electric Machine: The Greening of General Electric. 2005. *Economist*, December 10. Available at http://www.economist.com/node/5278338 (accessed January 13, 2011).

Lecocq, Franck, and Philippe Ambrosi. 2007. The Clean Development Mechanism: History, Status, and Prospects. *Review of Environmental Economics and Policy* 1 (1): 134–151.

Lee, David. 1996. Trading Pollution. In *Ozone Protection in the United States: Elements of Success*, ed. Elizabeth Cook, 31–38. Washington, DC: World Resources Institute.

Leggett, Jeremy. 2001. *The Carbon War: Global Warming and the End of the Oil Era*. New York: Routledge.

Levi-Faur, David. 2005. The Global Diffusion of Regulatory Capitalism. *Annals of the American Academy* 598 (1): 12–32.

Levy, David L., and Daniel Egan. 1998. Capital Contests: National and Transnational Channels of Corporate Influence on the Climate Change Negotiations. *Politics and Society* 26 (3): 337–361.

Levy, David L., and Daniel Egan. 2003. A Neo-Gramscian Approach to Corporate Political Strategy: Conflict and Accommodation in the Climate Change Negotiations. *Journal of Management Studies* 40 (4): 803–829.

Levy, David L., and Ans Kolk. 2002. Strategic Responses to Global Climate Change: Conflicting Pressures on Multinationals in the Oil Industry. *Business and Politics* 4 (3): 275–300.

Levy, David L., and Peter J. Newell. 2000. Oceans Apart? Business Responses to the Environment in Europe and North America. *Environment* 42 (9): 8–20.

Levy, David L., and Peter J. Newell. 2002. Business Strategy and International Environmental Governance: Toward a Neo-Gramscian Synthesis. *Global Environmental Politics* 2 (4): 84–101.

Levy, David L., and Peter J. Newell, eds. 2005. *The Business of Global Environmental Governance*. Cambridge, MA: MIT Press.

Levy, David L., and Aseem Prakash. 2003. Bargains Old and New: Multinational Corporations in Global Governance. *Business and Politics* 5 (2): 131–150.

Levy, David L., and Sandra Rothenburg. 1999. Corporate Strategy and Climate Change: Heterogeneity and Change in the Global Automobile Industry. BCSIA discussion paper E-99–13. Cambridge, MA: Kennedy School of Government.

Levy, David L., and Maureen Scully. 2007. The Institutional Entrepreneur as Modern Prince: The Strategic Face of Power in Contested Fields. *Organization Studies* 28 (7): 1–21.

Lieberman, Joe. 2007. Lieberman and Warner Introduce Bipartisan Climate Legislation. Press release. October 18. Available at http://lieberman.senate.gov/index.cfm/news-events/news/2007/10/lieberman-and-warner-introduce-bipartisan-climate-legislation (accessed March 20, 2009).

Lind, Nancy S., and Bernard I. Tamas. 2007. *Controversies of the George W. Bush Presidency*. Westport, CT: Greenwood Press.

Lisowski, Michael. 2002. Playing the Two-Level Game: US President's Decision to Repudiate the Kyoto Protocol. *Environmental Politics* 11 (4): 101–119.

Litfin, Karen T. 1995. Framing Science: Precautionary Discourse and the Ozone Treaties. *Millennium—Journal of International Studies* 24 (2): 251–277.

Llewellyn, John. 2007. *The Business of Climate Change: Challenges and Opportunities*. Lehman Brothers. Available at http://www.lehman.com/press/pdf_2007/TheBusinessOfClimateChange.pdf (accessed January 13, 2011).

Lohmann, Larry. 2006. *Carbon Trading: A Critical Conversation on Climate Change, Privatisation, and Power. Development Dialogue No. 48*. Uppsala: Dag Hammarskjöld Centre.

Loomis, Burdett A. 1986. Coalitions of Interest: Building Bridges in the Balkanized State. In *Interest Group Politics*, ed. Allan J. Cigler and Burdett A. Loomis, 258–274. Washington, DC: Congressional Quarterly Press.

Lowe, Ernest A., and Robert J. Harris. 1998. British Petroleum's Decision on Climate Change. *Corporate Environmental Strategy* 5 (2): 22–31.

Maeda, Risa. 2008. Japan Launches Voluntary CO_2 Market. Reuters, October 21. Available at http://www.reuters.com/article/idUSTRE49K1KR20081021 (accessed January 13, 2011).

Mahoney, Christine. 2007. Networking vs. Allying: The Decision of Interest Groups to Join Coalitions in the US and the EU. *Journal of European Public Policy* 14 (3): 366–383.

Makower, Joel, Ron Pernick, and Clint Wilder. 2008. *Clean Energy Trends 2008: March 2008*. Stamford, CT: Clean Edge.

Market Advisory Committee. 2007. *Recommendations for Designing a Greenhouse Gas Cap-and-Trade System for California*. Sacramento, CA: Market Advisory Committee to the California Air Resources Board.

Markussen, Peter, and Gert T. Svendsen. 2005. Industry Lobbying and the Political Economy of GHG Trade in the European Union. *Energy Policy* 33 (2): 245–255.

Marshall, Lord. 1998. *Economic Instruments and the Business Use of Energy.* Report. November. Available at http://www.hm-treasury.gov.uk/d/Economic Instruments.pdf (accessed March 20, 2009).

Matthews, Karine, and Matthew Paterson. 2005. Boom or Dust? The Economic Engine behind the Drive for Climate Change Policy. *Global Change, Peace, and Security* 17 (1): 59–75.

Mattli, Walter, and Tim Büthe. 2005. Global Private Governance: Lessons from a National Model of Setting Standards in Accounting. *Law and Contemporary Problems* 68 (3–4): 227–262.

Mattli, Walter, and Ngaire Woods. 2009. *The Politics of Global Regulation.* Princeton, NJ: Princeton University Press.

McAdam, Doug, John D. McCarthy, and Mayer N. Zald, eds. 1996. *Comparative Perspectives on Social Movements: Political Opportunities, Mobilizing Structures, and Cultural Framings.* Cambridge: Cambridge University Press.

McCarthy, John D., and Mayer N. Zald. 1978. Resource Mobilization and Social Movements: A Partial Theory. *American Journal of Sociology* 82 (6): 1212–1241.

McCauley, Kathy, Bruce Barron, and Morton Coleman. 2008. *Crossing the Aisle to Cleaner Air: How the Bipartisan "Project 88" Transformed Environmental Policy.* Pittsburgh: University of Pittsburgh, Institute of Politics.

Meckling, Jonas. 2008. Corporate Policy Preferences in the EU and the US: Emissions Trading as the Climate Compromise? *Carbon and Climate Law Review* 2 (2): 171–180.

Meckling, Jonas. 2011. The Globalization of Carbon Trading: Transnational Business Coalitions in Climate Politics. *Global Environmental Politics* 11 (2): 26–50.

Metcalf, Gilbert E. 2007. *A Proposal for a U.S. Carbon Tax Swap: An Equitable Tax Reform to Address Global Climate Change. The Hamilton Project.* Washington, DC: Brookings Institution.

Meyer, David S., and Catherine Corrigall-Brown. 2005. Coalitions and Political Context: U.S. Movements against Wars in Iraq. *Mobilization: International Journal* 10 (3): 327–344.

Mikler, John J. 2007. Varieties of Capitalism and the Auto Industry's Environmental Initiatives: National Institutional Explanations for Firms' Motivations. *Business and Politics* 9 (1): 1–38.

Milner, Helen V. 1997. *Interests, Institutions, and Information: Domestic Politics and International Relations.* Princeton, NJ: Princeton University Press.

Mintrom, Michael. 1997. Policy Entrepreneurs and the Diffusion of Innovation. *American Journal of Political Science* 41 (3): 738–770.

Mizruchi, Mark S. 1989. Similarity of Political Behavior among Large American Corporations. *American Journal of Sociology* 95 (2): 401–424.

Montgomery, W. David. 1972. Markets in Licenses and Efficient Pollution Control Programs. *Journal of Economic Theory* 5 (December): 395–418.

Morgan, Dan. 2001. Coal Scores with Wager on Bush. *Washington Post*, March 25, A5.

Mouawad, Jad. 2008. Industries Allied to Cap Carbon Differ on the Details. *New York Times*, June 2. Available at http://www.nytimes.com/2008/06/02/business/02trade.html (accessed January 13, 2011).

Mouawad, Jad. 2009. Businesses in U.S. Brace for New Rules on Emissions. *New York Times*, November 26. Available at http://www.nytimes.com/2009/11/26/business/energy-environment/26emissions.html (accessed January 13, 2011).

Mouawad, Jad, and Jeremy W. Peters. 2006. California Plan to Cut Gases Splits Industry. *New York Times*, September 1. Available at http://www.nytimes.com/2006/09/01/business/01energy.html (accessed January 13, 2011).

Murphy, David F., and Jem Bendell. 1997. *In the Company of Partners: Business, Environmental Groups, and Sustainable Development Post Rio*. Bristol: Policy Press.

Newell, Peter J. 2000. *Non-state Actors and the Global Politics of the Greenhouse*. Cambridge: Cambridge University Press.

Newell, Peter J. 2003. Globalization and the Governance of Biotechnology. *Global Environmental Politics* 3 (2): 56–71.

Newell, Peter J. 2008a. The Marketization of Global Environmental Governance: Manifestations and Implications. In *The Crisis of Global Environmental Governance*, ed. Jacob Park, Ken Conca, and Matthias Finger, 77–95. New York: Routledge.

Newell, Peter J. 2008b. The Political Economy of Global Environmental Governance. *Review of International Studies* 34 (3): 507–529.

Newell, Peter J., and David L. Levy. 2006. The Political Economy of the Firm in Global Environmental Governance. In *Global Corporate Power*, ed. Christopher May, 157–181. Boulder, CO: Lynne Rienner.

Newell, Peter J., and Matthew Paterson. 1998. A Climate for Business: Global Warming, the State, and Capital. *Review of International Political Economy* 5 (4): 679–703.

Newell, Peter J., and Matthew Paterson. 2010. *Climate Capitalism: Global Warming and the Transformation of the Global Economy*. Cambridge: Cambridge University Press.

Newton, Christopher. 2002. White House Shifted Carbon Dioxide Emissions Policy after Lobbyist's Letter. Associated Press, April 26.

Nicholson, Charles C. 2003. *Emissions Trading: A Market Instrument for Our Times*. London: Royal Society of Arts.

Nordhaus, William D. 2005. *Life after Kyoto: Alternative Approaches to Global Warming Policies*. Cambridge, MA: National Bureau of Economic Research.

Nye, Michael, and Susan Owens. 2008. Creating the UK Emission Trading Scheme: Motives and Symbolic Politics. *European Environment* 18 (1): 1–15.

Oberthür, Sebastian, and Hermann E. Ott. 1999. *The Kyoto Protocol: International Climate Policy for the Twenty-First Century*. Berlin: Springer.

O'Brien, Robert, Anne Marie Goetz, Jan Aart Scholte, and Marc Williams. 2003. *Contesting Global Governance: Multilateral Economic Institutions and Global Social Movements*. Cambridge: Cambridge University Press.

Offe, Claus, and Helmut Wiesenthal. 1980. Two Logics of Collective Action: Theoretical Notes on Social Class and Organizational Form. *Political Power and Social Theory* 1:67–115.

Oil and Gas Newsletter. 1997. *Oil and Gas Journal*, December 15, 2.

Oliver, Christine. 1991. Strategic Responses to Institutional Processes. *Academy of Management Review* 16 (1): 145–179.

Ougaard, Morten. 2006. Instituting the Power to Do Good? In *Global Corporate Power*, ed. Christopher May, 227–247. Boulder, CO: Lynne Rienner.

Park, Jacob, Ken Conca, and Matthias Finger. 2008. *The Crisis of Global Environmental Governance: Towards a New Political Economy of Sustainability*. London: Routledge.

Parry, Ian W. H., and William A. Pizer. 2007. Combating Global Warming. *Regulation* 30(3): 18–22.

Partnership for Climate Action. 2002. Common Elements among Advanced Greenhouse Gas Management Programs. Discussion paper. Washington, DC: Partnership for Climate Action.

Passell, Peter. 1988. Private Incentives as Pollution Curb. *New York Times*, October 19, D2.

Paterson, Matthew. 2001. Risky Business: Insurance Companies in Global Warming Politics. *Global Environmental Politics* 1 (4): 18–42.

Paterson, Matthew, and Peter J. Newell. 2008. Neo-Liberalism, Finance, and Legitimacy: Shaping Climate Capitalism. Paper presented at the Annual Convention of the International Studies Association, San Francisco, March 25–29.

Pattberg, Philipp H. 2007. *Private Institutions and Global Governance: The New Politics of Environmental Sustainability*. Cheltenham, UK: Edward Elgar.

Pattberg, Philipp H., and Johannes Stripple. 2008. Beyond the Public and Private Divide: Remapping Transnational Climate Governance in the Twenty-First Century. *International Environmental Agreement: Politics, Law, and Economics* 8 (4): 367–388.

Pauly, Louis W., and Simon Reich. 1997. National Structures and Multinational Corporate Behaviour: Enduring Differences in the Age of Globalization. *International Organization* 51 (1): 1–30.

Pedersen, Sigurd Lauge. 2001. The Danish CO_2 Emissions Trading Scheme. *Review of European Community and International Environmental Law* 9 (3): 223–231.

Petsonk, Annie, Daniel J. Dudek, and Joseph Goffman. 1998. *Market Mechanisms and Global Climate Change: An Analysis of Policy Instruments*. Washington, DC: Environmental Defense Fund.

Pew Center. 1999. ABB, Entergy, and Shell International Join Growing Corporate Effort to Address Climate Change. Press release. Available at <http://www.pewclimate.org/business/news?page=88> (accessed October 4, 2010).

Pew Center. 2006. State Action. Pew Center on Global Climate Change. Available at http://www.pewtrusts.org/uploadedFiles/wwwpewtrustsorg/Reports/Global_warming/pew_climate_101_stateaction.pdf (accessed January 13, 2011).

Philibert, Cédric, and Julia Reinaud. 2004. *Emissions Trading: Taking Stock and Looking Forward*. Paris: International Energy Agency.

Pigou, Arthur Cecil. 1920. *The Economics of Welfare*. London: Macmillan and Company.

Pinkse, Jonatan. 2006. *Business Responses to Global Climate Change*. Veenendaal, Netherlands: Universal Press.

Pinkse, Jonatan, and Ans Kolk. 2007. Multinational Corporations and Emissions Trading: Strategic Responses to New Institutional Constraints. *European Management Journal* 25 (6): 441–452.

Pinkse, Jonatan, and Ans Kolk. 2009. *International Business and Global Climate Change*. New York: Routledge.

Point Carbon. 2008. Carbon 2008: Post-2012 Is Now. Ed. Kjetil Roine, Endre Tvinnereim, and Henrik Hasselknippe. Copenhagen: Point Carbon.

Point Carbon. 2010. Carbon 2010: Return of the Sovereign. Ed. Endre Tvinnereim and Kjetil Roine. Oslo: Point Carbon.

Prakash, Aseem, and Matthew Potoski. 2006. *The Voluntary Environmentalists: Green Clubs, ISO 14001, and Voluntary Environmental Regulations*. Cambridge: Cambridge University Press.

Price, Richard. 1998. Reversing the Gun Sights: Transnational Civil Society Targets Land Mines. *International Organization* 52 (3): 613–644.

Pulver, Simone. 2004. *Power in the Public Sphere: The Battles between Oil Companies and Environmental Groups in the UN Climate Change Negotiations, 1991–2003*. Berkeley: University of California Press.

Pulver, Simone. 2005. Organising Business: Industry NGOs in the Climate Debates. In *The Business of Climate Change: Corporate Responses to Kyoto*, ed. Kathryn Begg, Frans van der Woerd, and David L. Levy, 47–60. Sheffield, UK: Greenleaf Publishing.

Pulver, Simone. 2007. Making Sense of Corporate Environmentalism. *Organization and Environment* 20 (1): 44–83.

Putnam, Robert. 1988. Diplomacy and Domestic Politics: The Logic of Two-Level Games. *International Organization* 42 (3): 427–460.

Rabe, Barry G. 2004. *Statehouse and Greenhouse*. Washington, DC: Brookings Institution Press.

Rabe, Barry G. 2006. Second-Generation Climate Policies in the American States: Proliferation, Diffusion, and Regionalization. *Issues in Governance Studies* 6 (August). Available at http://www.brookings.edu/~/media/Files/rc/papers/2006/08 energy_rabe.pdf (accessed January 13, 2011).

Rank, Michael. 1998. Shell Mulls Trade in CO_2 Emission Permits. *Climate News*, February 17.

Raustiala, Kal. 2002. States, NGOs, and International Environmental Institutions. *International Studies Quarterly* 41 (4): 719–740.

Raustiala, Kal, and David G. Victor. 2004. The Regime Complex for Plant Genetic Resources. *International Organization* 58 (2): 277–309.

Rees, Matthew, and Rainer Evers. 2000. Proposals for Emissions Trading in the United Kingdom. *Review of European Community and International Environmental Law* 9 (3): 232–238.

Reifschneider, Laura M. 2002. Global Industry Coalition. In *The Cartagena Protocol on Biosafety: Reconciling Trade in Biotechnology with Environment and Development?* ed. Christoph Bail, Robert Falkner, and Helen Marquard, 273–281. London: RIIA/Earthscan.

Reinhardt, Forest. 2001. *Global Climate Change and BP Amoco*. Case no. 9-700-106. Cambridge, MA: Harvard Business School.

Repetto and Dias. 2006. The True Picture? *Environmental Finance*, July–August, 44–45.

Revkin, Andrew C. 2001a. After Rejecting Climate Treaty, Bush Calls in Tutors to Give Courses and Help Set One. *New York Times*, April 28. Available at http://www.nytimes.com/2001/04/28/us/after-rejecting-climate-treaty-bush-calls-tutors-give-courses-help-set-one.html?src=pm (accessed January 13, 2011).

Revkin, Andrew C. 2001b. Bush's Shift Could Doom Air Pact, Some Say. *New York Times*, March 17. Available at http://query.nytimes.com/gst/fullpage.html?res=9800E1D8153DF934A25750C0A9679C8B63 (accessed January 13, 2011).

Revkin, Andrew C. 2001c. Bush Team under Attack on Emission Talks. *New York Times*, February 16. Available at http://www.nytimes.com/2001/02/16/us/bush-team-under-attack-on-emissions-talks.html (accessed January 13, 2011).

Revkin, Andrew C., and Neela Banerjee. 2001. Some Energy Executives Urge U.S. Shift on Global Warming. *New York Times*, August 1. Available at http://www.nytimes.com/2001/08/01/business/some-energy-executives-urge-us-shift-on-global-warming.html (accessed January 13, 2011).

RGGI. 2005. Memorandum of Understanding. Available at <http://www.rggi.org/docs/mou_12_20_05.pdf> (accessed October 4, 2010).

Rice, John G. 2007. *Addressing Climate Change: Views from Private Sector Panels. Subcommittee on Energy and Air Quality of the House Committee on Energy and Commerce*. Washington, DC: U.S. Congress.

Riker, William. 1986. *The Art of Political Manipulation*. New Haven, CT: Yale University Press.

Risse, Thomas, Stephen C. Ropp, and Kathryn Sikkink, eds. 1999. *The Power of Human Rights: International Norms and Domestic Change*. Cambridge: Cambridge University Press.

Rogers, Arthur. 2003. EU Parliament OKs Emissions Trading Plan. *International Environment Reporter* 26 (15): 714.

Rogers, Jim. 2005. *Business Actions Reducing Greenhouse Gas Emissions: Committee on Science, House of Representatives*. Washington, DC: Committee on Science.

Ronit, Karsten. 2007. *Global Public Policy: Business and the Countervailing Powers of Civil Society*. London: Routledge.

Ronit, Karsten, and Volker Schneider. 2000. *Private Organisations in Global Politics*. New York: Routledge.

Rowlands, Ian H. 2000. Beauty and the Beast? BP's and Exxon's Positions on Global Climate Change. *Environment and Planning C: Government and Policy* 18 (3): 339–354.

Rowlands, Ian H. 2001. Classical Theories of International Relations. In *International Relations and Global Climate Change*, ed. Urs Luterbacher and Detlef F. Sprinz, 43–66. Cambridge, MA: MIT Press.

Ruggie, John G. 1995. Peace in Our Time? Causality, Social Facts, and Narrative Knowing. *American Society of International Law.* Proceedings of the Eighty-ninth Annual Meeting: Structures of World Order, April 6, 1995 82:93–100.

Ruggie, John G. 1998. What Makes the World Hang Together? Neo-Utilitarianism and the Constructivist Challenge. *International Organization* 52 (4): 855–885.

Ruggie, John G. 2004. Reconstituting the Global Public Domain. *European Journal of International Relations* 10 (4): 499–531.

Ruggie, John G. 2007. Global Markets and Global Governance: The Prospects for Convergence. In *Global Liberalism and Political Order: Toward a New Grand Compromise?* ed. Steven Bernstein and Louis W. Pauly, 23–48. Albany: State University of New York Press.

Sabatier, Paul A. 1988. An Advocacy Coalition Framework of Policy Change and the Role of Policy-Oriented Learning Therein. *Policy Sciences* 21 (2/3): 129–168.

Sabatier, Paul A., and Hank C. Jenkins-Smith, eds. 1993. *Policy Change and Learning: An Advocacy Coalition Approach*. Boulder, CO: Westview Press.

Salisbury, Robert H., John P. Heinz, Edward O. Laumann, and Robert L. Nelson. 1987. Who Works with Whom? Interest Group Alliances and Opposition. *American Political Science Review* 81 (4): 1211–1234.

Sally, Razeen. 1994. Multinational Enterprises, Political Economy, and Institutional Theory: Domestic Embeddedness in the Context of Internationalization. *Review of International Political Economy* 1 (1): 161–193.

Sally, Razeen. 1995. *States and Markets: Multinational Enterprises in Institutional Competition*. London: Routledge.

Samson, Paul R. 1998. Non-State Actors and Environmental Assessment: North American Acid Rain and Global Climate Change. ENRP discussion paper E-98–10. Cambridge, MA: Kennedy School of Government.

Sandor, Richard. 1997. *Getting Started with a Pilot: The Rationale for a Limited-Scale Voluntary International Greenhousee Gas Emissions Trading Program. Energy and Natural Resources Committee*. Washington, DC: Centre Financial Products Ltd.

Sandor, Richard. 2001. The Case for Plurilateral Environmental Markets. *Environmental Finance,* September: 15.

Sandor, Richard. 2003. Institutional Innovation (Part II). *Environmental Finance*, May, 11.

Sandor, Richard, Michael Walsh, and Rafael L. Marques. 2002. Greenhouse-Gas-Trading Markets. *Philosophical Transactions of the Royal Society* 360 (1797): 1889–1900.

Schneider, Lambert. 2007. *Is the CDM Fulfilling Its Environmental and Sustainable Development Objectives? An Evaluation of the CDM and Options for Improvement: Report for the WWF*. Freiburg: Oeko-Institut.

Schneider, Mark, and Paul Teske. 1992. Toward a Theory of the Political Entrepreneur: Evidence from Local Government. *American Political Science Review* 86 (3): 737–747.

Schrope, Mark. 2001. A Change of Climate for Big Oil. *Nature* 411 (May 31): 516–518.

Schurman, Rachel. 2004. Fighting "Frankenfoods": Industry Opportunity Structures and the Efficacy of the Anti-Biotech Movement in Western Europe. *Social Problems* 51 (2): 243–268.

Scott, Dean. 2005. Bingaman Pledges Legislation in 2006 with Mechanism for Carbon Dioxide Trading. *Environment Reporter* 36 (48): 2539.

Scott, Dean. 2006. Environmental Groups Urge States, Cities to Avoid Voluntary Chicago Climate Effort. *Environment Reporter* 37 (33): 1715.

Scott, Dean. 2007a. Businesses Call on Congress to Act in 2007. *Environment Reporter* 38 (4): 176.

Scott, Dean. 2007b. Ruling May Accelerate Congressional Action toward Passage of Cap-and-Trade Legislation. *Environment Reporter* 38 (14).

Selin, Henrik, and Stacy D. VanDeveer. 2007. Political Science and Prediction: What's Next for U.S. Climate Change Policy? *Review of Policy Research* 24 (1): 1–27.

Selin, Henrik, and Stacy D. VanDeveer, eds. 2009. *Changing Climates in North American Politics: Institutions, Policymaking and Multilevel Governance*. Cambridge, MA: MIT Press.

Sell, Susan K. 2003. *Private Power, Public Law: The Globalization of Intellectual Property Rights*. Cambridge: Cambridge University Press.

Sell, Susan K., and Aseem Prakash. 2004. Using Ideas Strategically: The Contest between Business and NGO Networks in Intellectual Property Rights. *International Studies Quarterly* 48 (1): 143–175.

Shell. 1998. Profits and Principles: Does There Have to Be a Choice? *Shell Report*. Available at http://sustainabilityreport.shell.com/2008/servicepages/downloads/files/shell_report_1998.pdf (accessed January 13, 2011).

Simmons, Beth A. 1997. *Who Adjusts? Domestic Sources of Foreign Economic Policy during the Interwar Years*. Princeton, NJ: Princeton University Press.

Simmons, Beth A., Frank Dobbin, and Geoffrey Garrett. 2008. Introduction: The Diffusion of Liberalization. In *The Global Diffusion of Markets and Democracy*, ed. Beth A. Simons, Frank Dobbin, and Geoffrey Garrett, 1–63. Cambridge: Cambridge University Press.

Simmons, Beth A., and Zachary Elkins. 2004. The Globalization of Liberalization: Policy Diffusion in the International Political Economy. *American Political Science Review* 98 (1): 171–189.

Simmons, P. J. 1998. Learning to Live with NGOs. *Foreign Policy* 112:82–96.

Skidmore-Hess, Daniel. 1996. Business Conflict and Theories of the State. In *Business and the State in International Relations*, ed. Ronald W. Cox, 199–216. Boulder, CO: Westview Press.

Skjaerseth, Jon Birger, and Tora Skodvin. 2001. Climate Change and the Oil Industry: Common Problems, Different Strategies. *Global Environmental Politics* 1 (4): 43–64.

Skjaerseth, Jon Birger, and Tora Skodvin. 2003. *Climate Change and the Oil Industry: Common Problems, Varying Strategies*. Manchester: Manchester University Press.

Skjaerseth, Jon Birger, and Jorgen Wettestad. 2008. *EU Emissions Trading: Initiation, Decision-Making, and Implementation*. Hampshire, UK: Ashgate.

Slaughter, Anne-Marie. 2004. *A New World Order*. Princeton, NJ: Princeton University Press.

Snow, David A., and Robert D. Benford. 1988. Ideology, Frame Resonance, and Participant Mobilization. In *From Structure to Action: Social Movement Participation across Cultures*, ed. Bert Klandermans, Hanspeter Kriesi, and Sidney Tarrow, 197–217. Greenwich, CT: JAI Press.

Stauber, John C., and Sheldon Rampton. 1995. *Toxic Sludge Is Good for You*. Monroe, ME: Common Courage Press.

Stavins, Robert N., ed. 1988. *Project 88: Harnessing Market Forces to Protect the Environment*. Washington, DC, December.

Stavins, Robert N. 1998. What Can We Learn from the Grand Policy Experiment? Lessons from SO2 Allowance Trading. *Journal of Economic Perspectives* 12 (3): 69–88.

Stavins, Robert N. 2003. Market-Based Environmental Policies: What Can We Learn from U.S. Experience (and Related Research)? Discussion paper 03-43. Washington, DC: Resources for the Future.

Stavins, Robert N. 2007. *A U.S. Cap-and-Trade System to Address Global Climate Change: The Hamilton Project*. Washington, DC: Brookings Institution.

Streck, Charlotte. 2004. New Partnerships in Global Environmental Policy: The Clean Development Mechanism. *Journal of Environment and Development* 13 (3): 295–322.

Sullivan, John, and Pamela Najor. 2001. Two U.S. Senators Announce Intent to Draft Bill Creating Emissions Cap, Trade System. *International Environment Reporter* 24 (17): 700.

Svendsen, Gert T. 2002. *Lobbyism and CO_2 Trade in the EU*. Aarhus, Denmark: Aarhus School of Business.

Tangen, Kristian, and Henrik Hasselknippe. 2005. Converging Markets. *International Environmental Agreement: Politics, Law, and Economics* 5 (1): 47–64.

Tarrow, Sidney. 1998. *Power in Movement: Social Movements, Collective Action, and Politics*. New York: Cambridge University Press.

Tarrow, Sidney. 2005. *The New Transnational Activism*. Cambridge: Cambridge University Press.

Taylor, Rob, and James Thornhill. 2008. Australia Trading Scheme to Compensate Big Carbon Emitters. Reuters News. Available at http://www.reutersinteractive.com (accessed July 18, 2008). 16/07/08

Thacker, Strom C. 2000. *Big Business, the State, and Free Trade*. Cambridge: Cambridge University Press.

Thompson, Clive. 2008. A Green Coal Baron? *New York Times*, June 22. Available at http://www.nytimes.com/2008/06/22/magazine/22Rogers-t.html (accessed January 13, 2011).

Trading Thin Air. 2007. *Economist*, May 31. Available at http://www.economist.com/node/9217960?story_id=9217960 (accessed January 13, 2011).

UNCTAD. 1992. *Combating Global Warming: Study on a Global System of Tradeable Carbon Emission Entitlements*. New York: United Nations.

UNIPEDE/Eurelectric. 1999. *Greenhouse Gas and Electricity Trading Simulation*. Brussels: UNIPEDE/Eurelectric.

United Nations. 1995. Report of the Conference of the Parties on its First Session. Available at http://unfccc.int/resource/docs/cop1/07a01.pdf (accessed February 17, 2011).

United Nations. 1998. Kyoto Protocol to the United Nations Framework Convention on Climate Change. Available at http://unfccc.int/resource/docs/convkp/kpeng.pdf (accessed January 13, 2011).

United Nations Development Programme. 2002. *Human Development Report 2002: Deepening Democracy in a Fragmented World*. New York: Oxford University Press.

USCAP. 2007a. A Call for Action. Available at http://www.us-cap.org (accessed March 20, 2009).

USCAP. 2007b. Letter to Senators Lieberman and Warner. Available at http://www.us-cap.org (accessed March 20, 2009).

USCAP. 2007c. U.S. Climate Action Partnership Expands Membership to over 30 Organizations. Press release. Available at http://www.us-cap.org/media/release_071807.pdf (accessed October 4, 2010).

USCAP. 2009. USCAP Hails Waxman-Markey Climate Discussion Draft as Strong Start: U.C.A. Partnership. Available at http://www.us-cap.org/pdf/USCAP_Statement-on-Waxman-MarkeyDiscussion.pdf (accessed March 11, 2010).

U.S. Congress. 1997. Senate Resolution 98, 1st Sess. Available at http://www.senate.gov (accessed March 20, 2009).

U.S. Congress. 2003a. The Climate Stewardship Act of 2003. Available at http://frwebgate.access.gpo.gov/cgi-bin/getdoc.cgi?dbname=108_cong_bills&docid=f:s139is.txt.pdf (accessed October 4, 2010).

U.S. Congress. 2003b. The Science of Climate Change Senate Floor Statement by U.S. Senator James M. Inhofe. Available at http://inhofe.senate.gov/pressreleases/climate.htm (accessed October 4, 2010).

U.S. Congress. 2005. Sense of the Senate Resolution on Climate Change. Available at http://energy.senate.gov/public/index.cfm?FuseAction=PressReleases

.Detail&PressRelease_id=96bc510d-6c74-4a44-826d-9843a8ec9183&Month=6&Year=2005 (accessed October 4, 2010).

U. S. Department of State. 2007. U.S. Climate Action Report 2006. Available at <http://www.state.gov> (accessed August 21, 2008).

Useem, Bert, and Mayer N. Zald. 1982. From Pressure Group to Social Movement: Efforts to Promote Use of Nuclear Power. *Social Problems* 30 (2): 37–68.

U.S. National Science and Technology Council. 2005. National Acid Precipitation Assessment Program Report to Congress: An Integrated Assessment. Washington, DC. Available at http://www.esrl.noaa.gov/csd/aqrs/reports/napapreport05.pdf (accessed January 13, 2011).

Van der Veer, Jeroen. 2007. States Should Create a Climate of Change. *Financial Times* (North American ed.), January, 24, 15.

Victor, David G. 2001. *The Collapse of the Kyoto Protocol and the Struggle to Slow Global Warming*. Princeton, NJ: Princeton University Press.

Victor, David G. 2007. Fragmented Carbon Markets and Reluctant Nations: Implications for the Design of Effective Architectures. In *Architectures for Agreement: Addressing Global Climate Change in the Post-Kyoto World*, ed. Joseph Aldy and Robert N. Stavins, 133–160. New York: Cambridge University Press.

Victor, David G., and David Cullenward. 2007. Making Carbon Markets Work. *Scientific American* (September): 26.

Victor, David G., and Joshua C. House. 2004. A New Currency: Climate Change and Carbon Credits. *Harvard International Review* 26(2): 56–59.

Victor, David G., and Joshua C. House. 2006. BP's Emissions Trading Scheme. *Energy Policy* 34 (15): 2100–2112.

Victor, David G., Joshua C. House, and Sarah Joy. 2005. A Madisonian Approach to Climate Policy. *Science* 16 (5742): 1820–1821.

Vogel, David. 1986. *National Styles of Regulation: Environmental Policy in Great Britain and the United States*. Ithaca, NY: Cornell University Press.

Vogel, David. 1995. *Trading Up: Consumer and Environmental Regulation in a Global Economy*. Cambridge, MA: Harvard University Press.

von Bulow, Marisa. 2010. *Building Transnational Networks: Civil Society and the Politics of Trade in the Americas*. Cambridge: Cambridge University Press.

Wapner, Paul. 1995. Politics beyond the State: Environmental Activism and World Civic Politics. *World Politics* 47 (3): 311–341.

Wara, Michael W. 2007. Is the Global Carbon Market Working? *Nature* 445 (8): 595–596.

Wara, Michael W. 2009. Measuring the Clean Development Mechanism's Performance and Potential. *UCLA Law Review* 55 (6): 1759–1803.

Ware, Patricia. 2007a. International Coalition Formed in Lisbon to Establish Cap-and-Trade Carbon Market. *Environment Reporter* 38 (43).

Ware, Patricia. 2007b. Lieberman, Warner Release Draft Proposal for Industry Trading of Greenhouse Gases. *Environment Reporter* 38 (2): 1713.

Warrick, Joby, and Peter Baker. 1997. Clinton Details Global Warming Plan. *Washington Post,* October 23, A01.

Watanabe, Rie, and Guy Robinson. 2005. The European Union Emissions Trading Scheme. *Climate Policy* 5 (1): 10–14.

Weber, Steve. 1994. Origins of the European Bank for Reconstruction and Development. *International Organization* 48 (1): 1–38.

Wendt, Alexander. 1987. The Agent-Structure Problem in International Relations Theory. *International Organization* 41 (3): 335–370.

Wendt, Alexander. 1999. *Social Theory of International Politics.* New York: Cambridge University Press.

Werksman, Jacob. 1998. The Clean Development Mechanism: Unwrapping the Kyoto Surprise. *Review of European Community and International Environmental Law* 7 (2): 147–158.

Western Climate Initiative. 2007. History. Available at <http://www.westernclimateinitiative.org/history> (accessed October 4, 2010).

Whetzel, Carolyn. 2006. Schwarzenegger Backs Market Approach to Limit, Curb Emissions but Resists Cap. *Environment Reporter* 37 (15): 811.

Wilson, Graham K. 2003. Policy Transfer versus Varieties of Capitalism in Environmental Policy. Paper presented at the Internationalization of Regulatory Reform conference, University of California at Berkeley, April 25–26.

Wintour, Patrick. 2006. Blair Signs Climate Pact with Schwarzenegger. Guardian, August 1. Available at http://www.guardian.co.uk/environment/2006/aug/01/greenpolitics.usnews (accessed January 13, 2011).

Woll, Cornelia. 2009. *Firm Interests: How Governments Shape Business Lobbying on Global Trade.* Ithaca, NY: Cornell University Press.

Wynn, Gerard. 2007. Carbon Traders Bet on Bali Climate Talks' Success. Reuters, December 7.

Yamin, Farhana. 2005. *Climate Change and Carbon Markets: A Handbook of Emission Reduction Mechanisms.* London: Earthscan.

Yandle, Bruce. 1983. Baptists and Bootleggers: The Education of a Regulatory Economist. *Regulation* 7(3): 12–16.

Yandle, Bruce. 1998. Bootleggers, Baptists, and Global Warming. PERC Policy Series Number PS-14. Available at <http://www.perc.org/pdf/ps14.pdf> (accessed March 20, 2009).

Yandle, Bruce, and Stuart Buck. 2002. Bootleggers, Baptists, and the Global Warming Battle. *Harvard Environmental Law Review* 26 (1): 177–229.

Yi, Matthew. 2007. Proposed Air Board Leaders Pleases Dems. Governor's Nominee Advocates Need for Clean Air Regulations. *San Francisco Chronicle,* July 18. Available at http://articles.sfgate.com/2007-07-18/bay-area/17253711_1_market-systems-cap-and-trade-state-air-board (accessed January 13, 2011).

Young, Alasdair R. 2003. Political Transfer and "Trading Up"? Transatlantic Trade in Genetically Modified Food and U.S. Politics. *World Politics* 55 (4): 457–484.

Young, Oran R. 1989. *International Cooperation: Building Regimes for Natural Resources and the Environment*. Ithaca, NY: Cornell University Press.

Young, Oran R. 1997. Global Governance: Toward a Theory of Decentralized World Order. In *Global Governance: Drawing Insights from the Environmental Experience*, ed. Oran R. Young, 273–317. Cambridge, MA: MIT Press.

Zabarenko, Deborah. 2008. U.S. Carbon-Capping Climate Bill Dies in Senate. Reuters News, June 06. Available at http://www.reuters.com/article/idUSWAT 00961120080606 (accessed January 13, 2011).

Zald, Mayer N. 1996. Culture, Ideology, and Cultural Framing. In *Comparative Perspectives on Social Movements: Political Opportunities, Mobilizing Structures, and Cultural Framings*, ed. Douglas McAdam, John D. McCarthy, and Mayer N. Zald, 261–274. Cambridge: Cambridge University Press.

Zapfel, Peter, and Matti Vainio. 2002. Pathways to European Greenhouse Gas Emissions Trading. Available at <http://www.feem.it/userfiles/attach/Publication /NDL2002/NDL2002-085.pdf> (accessed October 4, 2010).

Zürn, Michael. 1998. The Rise of International Environmental Politics: A Review of Current Research. *World Politics* 50 (4): 617–649.

Index